John Kubi

Remediation of PCB Spills

Mitchell D. Erickson, Ph.D.
Environmental Research Division
Argonne National Laboratory
Argonne, Illinois

LEWIS PUBLISHERS
Boca Raton   Ann Arbor   London   Tokyo

Library of Congress Cataloging-in-Publication Data

Erickson, Mitchell D.
    Remediation of PCB Spills / author, Mitchell D. Erickson.
        p.   cm.
    Includes bibliographical references and index.
    ISBN 0-87371-945-X
    1. Polychlorinated biphenyls--Environmental aspects.   2. Chemical
spills--Environmental aspects.   I. Title
TD196.P65E75   1993
628.5'2--dc20   93-9635
                                                                                                        CIP

PRINTED IN THE UNITED STATES OF AMERICA
1 2 3 4 5 6 7 8 9 0

Printed on acid-free paper

Preface

This book is based on work done in support of the U.S. Environmental Protection Agency's efforts to control PCB spills under the Toxic Substances Control Act (PL 94-469) of 1976. The Office of Toxic Substances within EPA is charged with enforcement of this act, including the Section 6(e) provisions specifically regulating the manufacture, processing, distribution in commerce, or use of PCBs. One component of these regulations gives EPA latitude to grant exemptions to the ban if the manufacture, processing, distribution in commerce, or use is totally enclosed or will not present an unreasonable risk of injury to health or the environment. Thus, certain uses, including that in electrical equipment (capacitors and transformers) has persisted to this day. However, when an accident occurs during normal use, transport, or other situations, PCBs can be spilled into the environment and must be cleaned up according to 40 CRF 761.60(d). In order to provide guidance to enforcement, written guidelines for cleaning up PCB spills, with particular emphasis on the sampling design, sampling methods, and analytical methods, was requested.

This book is adapted from three EPA reports:

Verification of PCB Spill Cleanup by Sampling and Analysis

> prepared by Bruce A. Boomer, Mitchell D. Erickson, Stephen E. Swanson, Gary L. Kelso, David C. Cox, and Bradley D. Schultz

Field Manual for Grid Sampling of PCB Sites to Verify Cleanup

> prepared by Gary L. Kelso, Mitchell D. Erickson, and David C. Cox,

Cleanup of PCB Spills

> prepared by Mitchell D. Erickson, Bruce A. Boomer, and Gary L. Kelso

These three reports were intended primarily for the use of EPA enforcement personnel to provide sampling and analysis methods to determine compliance with EPA policy on the cleanup of PCB spills. There has been, of course, interest from those involved with PCB cleanup in the private sector to use these guidelines in planning their policies and cleanup activities. In order to make this information more broadly available, Lewis Publishers and I have adapted the information from the original reports and condensed it into a single, readily available volume.

This book is an account of work performed from the U.S. Environmental Protection Agency. It has not been reviewed nor approved by EPA. Specific questions regarding TSCA enforcement of PCB rules should be referred to current issue of 40 CFR 761, the EPA-headquarters Office of Toxic Substances (Washington DC), or the appropriate EPA Regional TSCA officials.

Mitchell D. Erickson
Argonne National Laboratory

D

About the Editor

Mitchell D. Erickson is a chemist at Argonne National Laboratory in Argonne, Illinois. He currently conducts and manages research projects that develop new technologies for environmental characterization, environmental restoration, and waste management, especially as applied to the cleanup of U.S. Department of Energy sites. His current research interests include development of on-site and *in-situ* chemical measurement technologies using GC, GC/MS, FTIR, sensors, and other techniques; development of fast, automated methods for determination of radionuclides in environmental samples; interfacing of chemical analyses with  sampling and monitoring equipment; integration of chemical, physical, and geological site characterization technologies; development of destruction processes for organic environmental contaminants; and improvement in analytical methods for PCBs and dioxins. He also consults on issues related to PCBs, dioxins, and other chemicals.

Dr. Erickson has conducted research on trace organic analysis in environmental and biological media and hazardous waste using GC, GC/MS, and GC/FTIR; developed methods for determination of PCBs; studied the combustion of PCBs under fire conditions and measured the chlorinated dibenzofurans and other compounds therefrom; and provided technical assistance to the U.S. EPA in permitting destruction processes for PCBs. He led the establishment, qualification, and operation of an EPA-qualified (Contract Laboratory Program) laboratory for determination of volatiles, semivolatiles, and pesticides/PCBs in environmental samples from DOE facilities. He has also studied levels of halogenated hydrocarbons and other organics in the environment and in human tissues and fluids and has developed methods for the determination of organic compounds, such as nitro-PNAs, in diesel exhaust.

Dr. Erickson has published another book (*Analytical Chemistry of PCBs, 1992, Lewis Publishers*), published over 60 papers in peer-reviewed journals, presented 70 papers at scientific meetings, and published 48 scientific reports. He is a member of the American Chemical Society, the American Society for Mass Spectrometry, the Society for Applied Spectroscopy, the Hazardous Materials Control Research Institute, and Sigma Xi (Scientific Research Society of North America). He received an A.B. degree in Chemistry (1972) from Grinnell College and a Ph.D. in Analytical Chemistry (1976) from the University of Iowa.

Acknowledgments

I wish to thank first and foremost my colleagues who contributed substantially to the work presented here: Bruce A. Boomer, Stephen E. Swanson, and Gary L. Kelso formerly or currently with Midwest Research Institute, Kansas City, MO and David C. Cox, and Bradley D. Schultz formerly of the Washington Consulting Group, Washington, DC.   The team has now disbanded, but I feel we made a substantial contribution to EPA's efforts at making sense of the TSCA PCB provisions, not only in the area discussed here, but also in chemical analysis, quality assurance, permit review, demonstration audits, determination of combustion products under fire conditions, and technical assistance.   Daniel T. Heggem (now of EPA-EMSL-Las Vegas) and John H. Smith of EPA's Office of Toxic Substances, Washington, DC were work assignment managers, valuable technical and policy resources, and trusted colleagues for the efforts described in this book.

Brian Lewis and Kathy Walters at Lewis Publishers have been helpful both in encouraging me to take on this project and in fulfilling their end of the bargain to produce, print, and distribute the book to you.

"Thank You" also to my family, especially my wife Colleen and my sons Adam, Carl, and Brendan, for their forbearance with my life as a scientist and PCB researcher.

Table of Contents

Part I

Verification of PCB Spill Cleanup by Sampling and Analysis

Bruce A. Boomer, Mitchell D. Erickson, Stephen E. Swanson, and
Gary L. Kelso, Midwest Research Institute
David C. Cox and **Bradley D. Schultz,** Washington Consulting Group

I. Introduction

The U.S. Environmental Protection Agency (EPA) under the authority of the Toxic Substances Control Act (TSCA) Section 6(e) and 40 CFR Section 761.60(d), has determined that polychlorinated biphenyl (PCB) spills must be controlled and cleaned up. The Office of Toxic Substances (OTS) has been requested to provide written guidelines for cleaning up PCB spills, with particular emphasis on the sampling design and sampling and analysis methods to be used for the cleanup of PCB spills.

This work assignment is divided into two phases. The reports of Phase I are presented in Draft Interim Report No. 1, Revision No. 1, "Cleanup of PCB Spills from Capacitors and Transformers,"by Gary L. Kelso, Mitchell D. Erickson, Bruce A. Boomer, Stephen E. Swanson, David C. Cox, and Bradley D. Schultz, submitted to EPA on January 9, 1985. Phase I consists of a review and technical evaluation of the available documentation on PCB spill cleanup, contacts with EPA Regional Offices and industry experts, and preparation of preliminary guidelines for the cleanup of PCB spills. The document was aimed at providing guidance in all aspects of spill cleanup for those organizations which do not already have working PCB spill cleanup programs.

Phase II, reported in this document, reviews the available sampling and analysis methodology for assessing the extent of spill cleanup by EPA enforcement officials. This report includes some of the information from the Phase I report, incorporates comments on the Phase I report and the general issue which were received at a working conference on February 26-27, 1985,

0-87371-945-X/93/0.00 + $.50

and addresses the issue from the perspective of developing legally defensible data for enforcement purposes.

This report, intended primarily for EPA enforcement personnel, outlines specific sampling and analysis methods to determine compliance with EPA policy on the cleanup of PCB spills. The sampling and analysis methods can be used to determine the residual levels of PCBs at a spill site following the completion of cleanup activities. Although the methodologies outlined in this document are applicable to PCB spills in general, specific incidents may require special efforts beyond the scope of this report. Future changes in EPA policy may affect some of the information presented in this document.

Following a summary of the report (Section II), Section III presents an overview of PCB spills and cleanup activities. The guidelines on sampling and analysis (Section IV) includes discussion of sampling design, sampling techniques, analysis, and quality assurance.

II. Summary

This report presents the results of Phase II of this work assignment. Phase I consisted of a review and technical evaluation of the available documentation on PCB spill cleanup, contacts with EPA Regional Offices, and preparation of preliminary guidelines for the cleanup of PCB spills. Phase II (this document) reviews the available sampling and analysis methodology for assessing the extent of spill cleanup by EPA enforcement officials. The report incorporates some of the information from the Phase I report and general issues received at a working conference on PCB spills.

The EPA has set reporting requirements for PCB spills and views PCB spills as improper disposal of PCBs. Cleanup activities have not been standardized since PCB spills are generally unique situations evaluated on a case-by-case basis by both the PCB owner (or his contractor) and the responsible EPA Regional Office. Components of the cleanup process may include protecting the health and safety of workers; reporting the spill; quick response/ securing the site; determination of materials spilled; cleanup procedures; proper disposal of removed PCB materials; and sampling and analysis. The level of action required is dependent on the amount of spilled liquid, PCB concentration, spill area and dispersion potential, and potential human exposure.

A sampling design is proposed for use by EPA enforcement staff in detecting residual PCB contamination above a designated limit after a spill site has been cleaned. The proposed design involves sampling on a hexagonal grid which is centered on the cleanup area and extends just beyond its boundaries. Guidance is provided for centering the design on the spill site, for staking out the sampling locations, and for taking possible obstacles into account. Additional samples can be collected at the discretion of the sampling crew.

Compositing strategies, in which several samples are pooled and analyzed together, are recommended for each of the three proposed designs. Since an enforcement finding of noncompliance must be legally defensible, the sampling design emphasizes the control of the false positive rate, the probability of concluding that PCBs are present above the allowable limit when, in fact, they are not.

Sampling and analysis techniques are described for PCB-contaminated solids (soil, sediment, etc.), water, oils, surface wipes, and vegetation. A number of analytical methods are referenced; appropriate enforcement methods were selected based on reliability. Since GC/ECD is highly reliable, widely used, and is included in many standard methods, it is a primary recommended method for most samples. Secondary methods may be useful for confirmatory analyses or for special situations when the primary method is not applicable.

Quality assurance (QA) must be applied throughout the entire monitoring program. Quality control (QC) measures, including protocols, certification and performance checks, procedural QC, sample QC, and sample custody as appropriate, should be stipulated in a QA plan.

III. Overview of PCB Spills and Cleanup Activities

A. Introduction to PCB Spills and Cleanup

The EPA has established requirements for reporting PCB spills based on the amount of material spilled and disposal requirements for the spilled PCBs and materials contaminated by the spill. Under TSCA regulations [40 CFR 761.30(a)(1)(iii) and 40 CFR 761.60d], PCB spills are viewed as improper disposal of PCBs. Although specific PCB cleanup requirements are not established in the TSCA regulations, each regional administrator is given authority by policy to enforce adequate clean-up of PCB spills to protect human health and the environment.

1. Current Trends

Due to regional variations in PCB spill policy and the lack of a national PCB cleanup policy, PCB cleanup activities have not been standardized. Individual companies owning PCB equipment and contract cleanup companies have developed their own procedures and policies for PCB cleanup activities keyed to satisfying the requirements of the appropriate EPA Regional Office. In addition, the EPA Regional Offices typically have provided suggestions for companies unfamiliar with PCB cleanup.

PCB spills are generally viewed as unique situations to be evaluated on a case-by-case basis by both the PCB owner (or his contractor) and the EPA Regional Office. However, a general framework is often used to approach the problem. Most cleanup activities involve quick response, removal or cleaning of suspected contaminated material, and post-cleanup sampling to document adequate cleanup. Major considerations involved in the cleanup process include minimizing environmental dispersion, minimizing any present or future human exposure to PCBs, protecting the health and safety of the cleanup crew, and properly disposing contaminated materials.

In general, the involvement of EPA Regional Offices is limited to phone conversations often including a follow-up call to receive the analytical results of the post-cleanup sampling. If the EPA representative is not satisfied with the reported data, additional documentation, sampling and analysis, or cleanup (followed by further sampling and analysis) may be requested.

In cases of special concern (e.g., large spills), EPA Regional Offices may work more closely with the PCB owner or contractor in planning the cleanup, sampling and analysis activities, and on-site inspections.

2. Limitations of This Overview

The general discussion in this chapter refers to the procedures, policy, and considerations that seem to be widely used at present by PCB owners and spill cleanup contractors in meeting the requirements of the EPA Regional Offices. The activities described do not involve EPA regulations or policy except where indicated, since the EPA has not established requirements on PCB cleanup procedures.

Table 1 categorizes PCB spills into approximate levels of action for PCB spill cleanup based on concern. Potential environmental problems increase with increases in PCB concentrations, amount of spilled liquid, spill area and dispersion potential, and potential human exposure. The three spill types presented in Table 1 are based on very rough estimates. "Severity" in one key item such as human exposure could raise a spill to a Type 3 (i.e., requiring special attention). On the other hand a spill of a large volume of liquid may be considered a Type 2 spill due to a relatively low concentration of PCBs. The three categories are only approximate and are intended to demonstrate the flexibility needed in responding to PCB spills. EPA Regional Offices should provide guidance on spill cleanup activities whenever questions develop.

Table 1. Approximate Levels of Action for PCB Spill Cleanup Based on Concern

	Categories of increasing concern		
	Type 1	Type 2	Type 3
Approximate gallons of spilled liquid	< 1	> 1	> 5
Area of spill (sq ft)	< 125	250 (avg.)	> 1,000
PCB concentration in spilled liquid (ppm)	< 500	≥ 50	Variable or high
Types of spilled liquid	Mineral oil (or variable)	Variable	Variable, Askarel
Exposure scenario	Various	Various	Special concern for high exposure situations

Notes: • **Type 1** spill is usually not reported.
   • **Type 2** spill is reported and discussed in this chapter
   • **Type 3** spill is not discussed in this chapter and may require special EPA assistance
   • "Severity" in one key item may raise the spill to a higher risk category.

The situations described in this chapter are limited to recent PCB spills of similar magnitude to the reported spills associated with PCB oil transformers and capacitors (i.e., Type 2 in Table 1). Unusually severe spill incidents (Type 3 in Table 1) involving large volumes of PCBs, a large spill area, a high probability of significant human exposure, and/or severe environmental or transportation scenarios may require special considerations, beyond the scope of this discussion.

All spills from regulated equipment are typically subject to the detail of effort outlined in this chapter. Although cleanup of smaller spills (Type 1 in Table 1) is required if the concentration of PCBs in the spilled material is 50 ppm or greater, the spill and the cleanup activities normally are not reported to EPA.

Future changes in EPA policy may invalidate some of the discussions appearing in this chapter. For example, if EPA adopts any type of formal categorization scheme for PCB spills, some of the assumptions made in this chapter may become inappropriate.

B. Components of the Cleanup Process

1. Health and Safety

Protection of the health and safety of the clean-up crew during the PCB cleanup operation is an important concern. References discussing health and safety considerations relevant to some PCB spill incidents include NIOSH Criteria for A Recommended Standard for Exposure to Polychlorinated Biphenyls (PCBs) (1977c) and Health Hazards and Evaluation Report No. 80-85-745 (NIOSH 1980). The appropriate level of health and safety protection is dependent upon the specifics of the spill.

2. Reporting the Spill

If the regulatory limits are exceeded, the spill must be reported to Federal, State, and local authorities as applicable. Under EPA regulations [Fed. Reg. 50:13456-13475], spills over 10 lb must be reported to The National Response Center. The toll free phone number is (800) 424-8802.

3. Quick Response/Securing the Site

Quick response is desirable to mitigate the dispersion of the spilled material and to secure the site. Federal regulations require that cleanup actions commence within 48 hr of discovery of a spill [40 CFR 761.30(a)(1) (iii)]. More rapid response is highly preferable.

A quick response allows removal or cleaning of the PCB-contaminated material before it is dispersed by wind, rain, seepage, and other natural causes or by humans or animals. In securing the site, the cleanup crew determines the spill boundaries, prevents unauthorized access to the spill site, and notifies all parties involved.

The methods used to secure the site will vary on a case-by-case basis, depending on the specific circumstances. The extent of the spill is usually determined by visual inspection with the addition of a buffer area that may include PCBs finely dispersed from splattering. Evaluating the extent of the spill involves considerable judgment, including consideration of the cause of the spill, weather conditions, and specifics of the site.

Field analysis kits may aid the crew in determining the extent of the spill in some instances. The field kits, when used properly, can serve as a screening tool. The need for quick response has limited the usefulness of the more accurate field analytical techniques such as field gas chromatography. Practical problems associated with availability of the equipment and trained staff, set-up time, and cost have limited the use of such techniques at this time.

4. Determination of Materials Spilled/Cleanup Plan

After securing the site, the response crew will either (a) immediately proceed with the cleanup operation, or (b) identify the materials spilled and formulate an appropriate cleanup plan. A suitable cleanup plan can be developed by identifying the type of PCB material (i.e., mineral oil, PCB oil, Askarel) and considering such factors as the volume spilled, area of the spill, and site characteristics.

Based on reasoning similar to Table 1, the crew leader can determine the necessary level of effort in accordance with the policy of the PCB owner and the EPA Regional Office. He can determine if additional guidance is needed, plan the sampling and analysis, and make other decisions related to the level of effort and procedures needed.

5. Cleanup Procedures

The cleanup procedure may include, but may not necessarily be limited to, the following activities:

- Removal or repair of failed/damaged PCB equipment,
- Physical removal of contaminated vegetation;
- Physical removal of contaminated soils, liquids, etc.,
- Decontamination or physical removal (as appropriate) of contaminated surfaces, and
- Decontamination or removal of all equipment potentially contaminated during the cleanup procedures.
- Encapsulation may be employed only with EPA approval.

The specific procedures used in a cleanup are selected by the PCB owner or the cleanup contractor. Key considerations include removal of PCBs from the site to achieve the standards required by the EPA region, company, or other applicable control authority; avoidance of unintentional cross contamination or dispersion of PCBs from workers' shoes, contaminated equipment, spilled cleaning solvents, rags, and other sources; and protection of workers' health.

The cleanup crew shall make every possible effort to keep the spilled PCBs out of sewers and waterways. If this has already occurred, the crew needs to contact the local authorities. Water is never used for cleaning equipment or the spill site.

A simple PCB spill cleanup may involve the removal of the leaking equipment, removal of contaminated sod and soil by shovel, cleaning pavement with an absorbent material and solvents, and decontamination or disposal of the workers' equipment (shovels, shoes, gloves, rags, plastic sheets, etc.). More complicated situations may include decontamination of cars, fences, buildings, trees and shrubs, electrical equipment, or water (in pools or bodies of water).

In some cases, adequate decontamination of surfaces (pavements, walls, etc.) may not be possible. An alternate to physical removal of the surface material is encapsulation of the contaminated area under a coating impervious to PCBs. (EPA approval would be required.)

6. Proper Disposal of Removed PCB Materials

All PCB-contaminated materials removed from the spill site, must be shipped and disposed in accordance with relevant Federal, State, and local regulations. TSCA Regulations [40 CFR 761.60] outline the requirements for the disposal of PCBs, PCB articles, and PCB containers in an incinerator, high efficiency boiler, chemical waste landfill, or an approved alternative method. Facility requirements for incineration and chemical waste landfills are presented in 40 CFR 761.70 and 40 CFR 761.75, respectively. Applicable Department of Transportation regulations are listed in 49 CFR 172.101.

7. Sampling and Analysis

Although sampling and analysis will be discussed in detail in Chapter IV, this discussion gives an overview of applicable considerations and current practice. Sampling and analysis may not always be needed (especially for the spills described as Type 1 in Table 1), but enforcement authorities or property owners may ask for proof that the spill site has been adequately decontaminated. This can be accomplished by taking a number of samples representative of the area contaminated by the spill. Samples should represent the full extent of the spill, both horizontal and vertical, as well as the types of materials in the spill area (soil, surfaces, water, etc.).

Sampling design and technique as well as sample handling and preservation should incorporate acceptable procedures for each matrix to be sampled and concern for the adequacy and accuracy for the samples in the final analysis.

Analysis of the samples for PCB content should be performed by trained personnel using acceptable procedures with due consideration of quality assurance and quality control.

Further discussion of sampling and analysis (applicable to EPA enforcement activities) appears in Chapter IV.

8. Remedial Action

If the analysis results indicate the cleanup was not in compliance with designated cleanup levels, additional cleanup is needed. Additional sampling can pinpoint the location of remaining contaminated areas if the original sampling plan was not designed to identify contaminated sub-areas within the spill site. If additional cleanup is needed, the cleanup crew will continue as before, removing more material or cleaning surfaces more thoroughly. Remedial action will be followed by additional sampling and analysis to verify the adequacy of the cleanup.

9. Site Restoration

This is not addressed under TSCA and is a matter to be settled between the company responsible for the PCB spill and the property owner.

10. Records

Although there are no TSCA requirements for records of PCB cleanup activities except for documentation of PCBs stored or transported for disposal [40 CFR 761.80(a)], the PCB owner should keep records of the spill cleanup in case of future questions or concern. Relevant information may include dates, a description of the activities, records of shipment and disposal of PCB-contaminated materials, and a report of collected samples and results of analysis.

11. Miscellaneous Considerations

a. Expeditious and effective action are desired throughout the cleanup process to minimize the concern of the public, especially residents near the site or individuals with a special interest in the site. Likewise, speed and effectiveness in the cleanup may prevent any future concern or action related to the PCB spill.

b. Education and training of the spill response crews and responsible staff members is a constant concern. The employees need sufficient training to make proper judgements and to know when additional assistance or guidance is needed.

IV. Guidelines on Sampling and Analysis

Reliable analytical measurements of environmental samples are an essential ingredient of sound decisions for safeguarding public health and improving the quality of the environment. Effective enforcement monitoring should follow the general operational model for conducting analytical measurements of environmental samples, including: planning, quality assurance/quality control, verification and validation, precision and accuracy, sampling, measurements, documentation, and reporting. Although many options are available when analyzing environmental samples, differing degrees of reliability, dictated by the objectives, time, and resources available, influence the protocol chosen for enforcement monitoring. The following section outlines the factors critically influencing the outcome and reliability of enforcement monitoring of PCB spill cleanup.

A. Sampling Design

This section presents a sampling scheme, for use by EPA enforcement staff, for detecting residual PCB contamination above a limit designated by EPA-OPTS after the site has been cleaned up. Two types of error traceable to sampling and analysis are possible. The first is <u>false positive</u>, i.e., concluding that PCBs are present at levels above the allowable limit when, in fact, they are not. The false positive rate for the present situation should be low, because an enforcement finding of noncompliance must be legally defensible; that is, a violator must not Se able to claim that the sampling results could easily have been obtained by chance alone. Moreover, all sampling designs used must be documented or referenced.

The second type of error possible is a <u>false negative</u>, i.e., failure to detect the presence of PCB levels above the allowable limit. The false negative rate will depend on the size of the contaminated area and on the level of contamination. For large areas contaminated at levels well above the allowable limit, the false negative rate must, of course, be low to ensure that the site is brought into compliance. The false negative rate can increase as the area or level of contamination decrease.

1. Proposed Sampling Design

In practice, the contaminated area from a spill will be irregular in shape. In order to standardize sample design and layout in the field, and to protect against underestimation of the spill area by the cleanup crew, sampling within a circular area surrounding the contaminated area is proposed. Guidance on choosing the center and radius of the circle, as well as the number of sample points to be used is provided in Section 2 below.

The detection problem was modeled as follows: try to detect a circular area of uniform residual contamination whose center is randomly placed within the sampling circle. Figure 1 illustrates the model. The figure depicts a sampling circle of 10 ft centered on a utility pole (site of the spill). After cleanup, a residually contaminated circle remains. However, in choosing locations at which to sample, the sampler has no knowledge of either the location of the circle or the level of contamination. This lack of knowledge was modeled by treating the sampling locations as fixed and the center of the contaminated circle as a randomly located point in the circle of radius 10 ft. The implicit assumption that residual contamination is equally likely to be present anywhere within the sampling area is reasonable, at least as a first

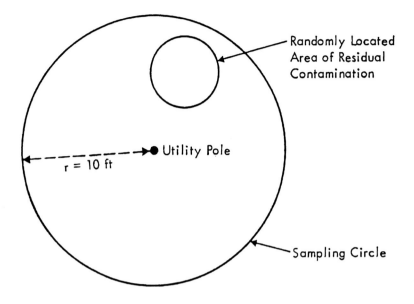

Figure 1. Randomly located area of residual contamination within the sampling circle.

approximation (Lingle 1985). This is because more effort is likely to have been expended in cleaning up the areas which were obviously highly contaminated.

Two general types of design are possible for this detection problem: grid designs and random designs. Random designs have two disadvantages compared to grid designs for this application. First, random designs are more difficult to implement in the field, since the sampling crew must be trained to generate random locations onsite, and since the resulting pattern is irregular. Second, grid designs are more efficient for this type of problem than random designs. A grid design is certain to detect a sufficiently large contaminated area while some random designs are not. For example, the suggested design with a sample size of 19 has a 100% chance to detect a contaminated area of radius 2.8 ft within a sampling circle of radius 10 ft. By contrast, a design based on a simple random sample of 19 points has only a 79% chance of detecting such an area.

Therefore, a grid design is proposed. A hexagonal grid based on equilateral triangles has two advantages for this problem. First, such a grid minimizes the circular area certain to be detected (among all grids with the same number of points covering the same area). Second, some previous experience (Mason 1982; Matern 1960) suggests that the hexagonal grid performs well for certain soil sampling problems. The hexagonal grid may, at first sight, appear to be complicated to lay out in the field. Guidance is provided in Section 2 below and shows that the hexagonal grid is quite practical in the field and is not significantly more difficult to deploy than other types of grid.

The smallest hexagonal grid has 7 points, the next 19 points, the third 37 points as shown in Figures 2 through 4. In general, the grid has $3n^2 + 3n + 1$ points. To completely specify a hexagonal grid, the distance between adjacent points, s, must be determined. The distance s was chosen to minimize, as far as possible, the size of the residual contaminated circle which is certain to be sampled. Values of s so chosen, together with number of sampling points and radius of smallest circle certain to be sampled, are shown in Table 2. For example, the grid spacing for a circle of radius 20 ft for the 7-point design is $s = (0.87)(20) = 17.4$ ft. For a given size circle, the more points on the grid, the smaller the residual contamination area which can be detected with a given probability.

For cases in which the configuration of the contaminated area is very different from a circle (e.g., an extremely elongated ellipse), the sampling circle may be a poor approximation to the contaminated area, and a moderate-to-large percentage of the sampling points may fall outside the contaminated area. If the sampler is certain that there is no contamination outside the cleanup area, then it is permissible to disregard those sampling points falling outside the cleanup area.

Table 2. Parameters of Hexagonal Sampling Designs for a Sampling Circle of Radius r Feet

No. of points	Distance between adjacent points, s (ft)	Radius of smallest circle certain to be sampled
7	0.87r	0.5r
19	0.48r	0.28r
37	0.3r	0.19r

It is still good practice to collect samples from these outlying points even if they are not ever analyzed because the cost of returning to the site to perform sampling activity again is much greater than the cost of incremental sampling performed while still onsite. If sampling points outside the contaminated area are ignored, and if it is a certainty that there is no contamination outside this area, the absolute detection capability of the sampling scheme is unaffected. For example, the chance of detecting a 5 sq ft area of contamination within the restricted sampling area is the same as it would be if the contaminated area comprised the entire sampling circle.

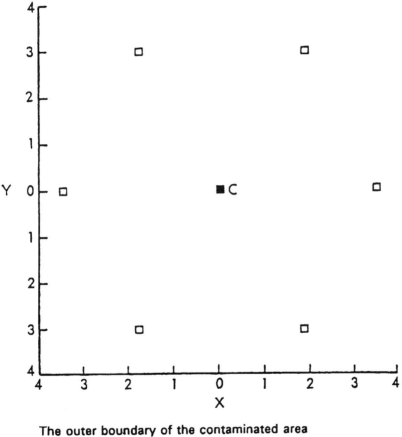

The outer boundary of the contaminated area is assumed to be 4 feet from the center (C) of the spill site.

Figure 2. Location of sampling points in a 7-point grid.

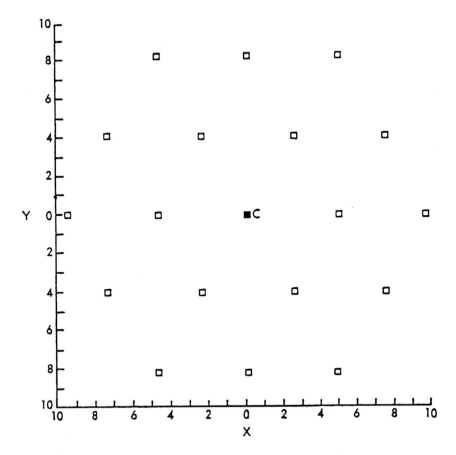

The outer boundary of the contaminated area is assumed to be
10 feet from the center (C) of the spill site.

Figure 3. Location of sampling points in a 19-point grid.

The first three hexagonal designs are shown in Figures 2 to 4, for a sampling circle radius of r = 10 ft. The choice of sample size depends on the cost of analyzing each sample and the reliability of detection desired for various residually contaminated areas. Subsection 2 below provides some suggested sample sizes for different spill areas, based on the distribution of spill areas provided by the Utility Solid Waste Activities Group (USWAG 1984; Lingle 1985).

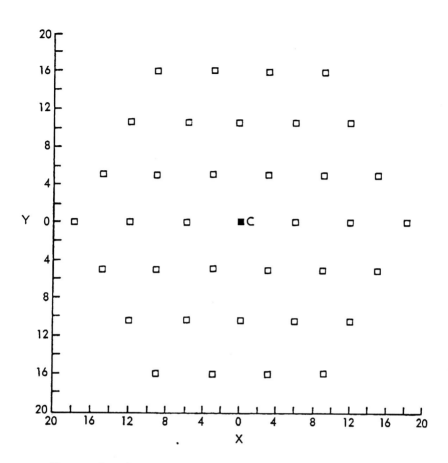

The outer boundary of the contaminated area is assumed to be 20 feet from the center (C) of the spill site.

Figure 4. Location of sampling points in a 37-point grid.

2. *Sample Size and Design Layout in the Field*

a. Sample Size

The distribution of cleanup areas for PCB capacitor spill sites, based on data collected by USWAG (1984; Lingle 1985) is shown in Table 3. The smallest spill recorded in the USWAG database is 5 ft^2 the largest 1,700 ft^2. The median cleanup area is 100 ft, the mean 249 ft^2; the wide discrepancy between the mean and the median reflects the presence of a small percentage of relatively large spills in the database.

Recommended sample sizes are given in Table 4. Several considerations were involved in arriving at these recommendations. First, the maximum number of samples recommended for the largest spills is 37, in recognition of practical constraints on the number of samples that can be taken. Even so, it is important to note that not all samples collected will need to be analyzed. The calculations in Section 5 below show that, even for the 37 sample case, no more than 8 analyses will usually be required to reach a decision. Since the cost of chemical analyses is a substantial component of sampling and analysis costs, even the 37-sample case should not, therefore, be prohibitively expensive. Second, the typical spill will require 19 samples. Small spills, with sampling radius no greater than 4 ft, will have 7 samples, while the largest spills, with sampling radius 11.3 ft and up, will require 37 samples. It should be noted that only capacitor spills are represented in Table 3. Transformer spills, however, would be expected to be generally smaller than capacitor spills because energetic releases are less likely from transformers. Thus, one would expect the smaller sample sizes to be relatively more likely for transformer spills than capacitor spills.

Table 3. Distribution of PCB Capacitor Spill Cleanup Areas Based on 80 Cases

Cleanup area (ft^2)	Percent of cases
≤ 50	32.5
51-100	18.8
101-200	15.0
201-300	12.5
301-400	3.8
401-700	7.5
701-1,300	8.8
≥ 1,300	1.3

Source: Lingle 1985.

Table 4. Recommended Sample Sizes

Sampling area (ft^2)	Radius of sampling circle (ft)	Percent of PCB capacitor spills	Sample size
≤ 50	≤ 4	32.5	7
51-400	4-11.3	50.0	19
> 400	> 11.3	17.5	37

The final consideration in recommending sample sizes was to achieve roughly comparable detection capability for different size spills. The radius of the smallest contaminated circle certain to be sampled at least once by the sampling scheme is used for comparative purposes (see Table 2). Table 5 presents some calculations of this quantity. The absolute detection capability of the sampling scheme is seen to be relatively constant for different spill sizes. This means that a given area of residual contamination is about as likely to be detected in any sized spill.

Table 5. Detection Capability of the Recommended Sampling Schemes

Sampling area (ft²)	Radius (ft)	Sample size	Radius of smallest circle to be sampled (ft)
50	4.0	7	2.0
150	6.9	19	1.9
400	11.3	19	3.2
875	16.7	37	3.2

b. Design Layout in the Field

Figure 5 presents a typical illustration of design layout in the field. The first step is to determine the boundaries of the original cleanup area (from records of the cleanup). Next, find the center and radius of the sampling circle which is to be drawn surrounding the cleanup area. The following approach is recommended:

(a) Draw the longest dimension, L_1, of the spill area.
(b) Determine the midpoint, P, of L_1.
(c) Draw a second dimension, L_2, through P perpendicular to L_1.
(d) The midpoint, C, of L_2 is the required center.
(e) The distance from C to the extremes of L_1 is the required radius, r.

Figure 5 shows an example of the procedure; Figure 6 demonstrates how the center is determined for several spill shapes. Even if the center determined is slightly off, the sampling design will not be adversely affected.

Once the sampling radius, r, has been found, the sample size can be selected based on Table 4.

Example: Suppose r = 5 ft. From Table 4, a sample size of 19 should be used.

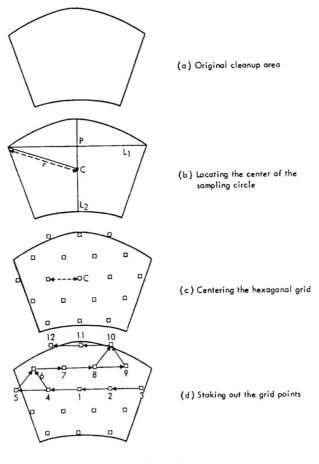

(a) Original cleanup area

(b) Locating the center of the sampling circle

(c) Centering the hexagonal grid

(d) Staking out the grid points

Figure 5

Having selected the sample size, the grid spacing can be calculated from Table 2.

Example: (continued): For a 19-point design with radius $r = 5$, the grid spacing is $s = 0.48r = (0.48)(5) = 2.4$ ft.

The procedure for laying out a 19 point design is as follows. The first sampling location is the center C of the sampling circle, as shown in Figure 5. Next, draw a diameter through C and stake out locations 2 through 5 on it as shown; adjacent locations are a distance s apart. The orientation of the diameter (for example east-west) used is not important; it may be chosen at

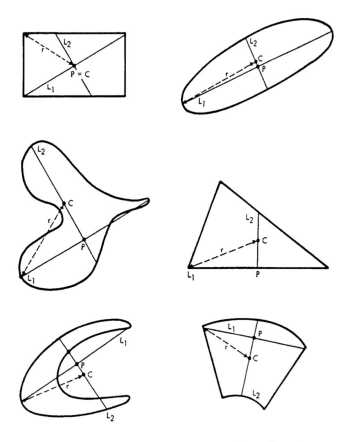

Figure 6. Locating the center and sampling circle radius of an irregularly shaped spill area.

random or for the convenience of the samplers. The next 4 locations, Nos. 6-9, are laid out parallel to the first row, again a distance s apart. The only difficulty is in locating the starting point, No. 6, for this row. To accomplish this the sampler needs two pieces of rope (or surveyor's chain, or equivalent measuring device) of length s. Attach one piece of rope to the stake at each location 4 and 5. Draw the ropes taut horizontally until they touch at location 6. Once the second row is laid out, the third and final row of 3 locations in the top half of the design is found similarly, starting with number 10. In the same way, the bottom half of the design is staked out. The 7-point or 37-point designs are laid out in an analogous fashion.

Once the sampling locations are staked out the actual samples can be collected. In the example in Figure 5, three of the sampling locations fall outside the original cleanup area. Samples should be taken at these points, to detect contamination beyond the original cleanup boundaries. This verifies that the original spill boundaries were accurately assessed. However, if the sampler is certain that there is no contamination outside the original cleanup area, then it is permissible to disregard those sampling points falling outside the cleanup area. It is still good practice, however, to collect such samples even if they are not ever analyzed because the cost of returning to the site to sample again is much greater than the cost of incremental sampling performed while still onsite. As indicated before, ignoring the sampling points outside the original cleanup area does not affect the absolute detection capability of the sample scheme.

In practice, various obstacles may be encountered in laying out the sampling grid. Many "obstacles" can be handled by taking a different type of sample, e.g., if a fire hydrant is located at a point in a sampling grid otherwise consisting of soil samples, then a <u>wipe</u> sample should be taken at the hydrant, rather than taking a sample of nearby soil. The obstacle most likely to be encountered is a vertical surface such as a wall. To determine the sampling location on such a surface, draw taut the ropes (chains) of length s attached to two nearby stakes and find the point on the vertical surface where their common ends touch. See Figure 7 for an illustration of the procedure. if more samples from the vertical surface are called for, the same principle may be applied, always using the last two points located to find the next one.

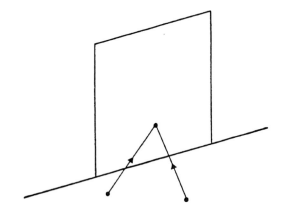

Figure 7. Location of a sampling point on a vertical
surface.

3. Judgmental Sampling

The inspector or sampling crew may use best judgement to collect samples wherever residual PCB contamination is suspected. These samples are in addition to those collected from the sampling grid. Examples of extra sampling points include suspicious stains outside the designated spill area, cracks or crevices, and any other area where the inspector suspects inadequate cleanup.

4. Compositing Strategy for Analysis of Samples

Once the samples have been collected at a site, the goal of the analysis effort is to determine whether <u>at least one</u> sample has a PCB concentration above the allowable limit. This sampling plan assumes the entire spill area will be recleaned if a single sample contaminated above the limit is found. Thus, it is not important to determine precisely which samples are contaminated or even exactly how many. This means that the cost of analysis can be substantially reduced by employing <u>compositing</u> strategies, in which groups of samples are thoroughly mixed and evaluated in a single analysis. If the PCB level in the composite is sufficiently <u>high</u>, one can conclude that a contaminated sample is present; if the level is <u>low</u> enough, all individual samples are clean. For intermediate levels, the samples from which the composite was constructed must be analyzed individually to make a determination. Thus, the number of analyses needed is greatly reduced in the presence of very high levels of contamination in a few samples or in the presence of very low levels in most samples.

For purposes of this discussion, assume that the maximum allowable PCB concentration in a single soil sample is 10 ppm. The calculations can easily be adapted for a different level or for different types of samples. Based on review of the available precision and accuracy data (Erickson 1985), method performance of 80% accuracy and 30% relative standard deviation should be attainable for soil concentrations above 1 ppm.

To protect against false positive findings due to analytical error, the measured PCB level in a single sample must exceed some cutoff greater than 10 ppm for a finding of contamination. Assume that a 0.5% false positive rate for a single sample is desired. As will be shown later, this single sample false positive rate controls the overall false positive rate of the sampling schemes to acceptable levels. Then, using standard statistical techniques, the cutoff level for a single sample is

$$(0.8)(10) + (2.576)(0.3)(0.8)(10) = 14.2 \text{ ppm,}$$

where 0.8(80%) represents the accuracy of the analytical method, 10 ppm is the allowable limit for a single sample, 2.576 is a coefficient from the standard normal distribution, and 0.3(30%) is the relative standard deviation of the analytical method. Thus, if the <u>measured</u> level in a single sample is 14.2 ppm or greater, one can be 99.5% sure that the <u>true</u> level is 10 ppm or greater.

Now suppose that a composite of, say, 7 samples is analyzed. The true PCB level in the composite (assuming perfect mixing) is simply the average of the 7 levels of the individual samples. Let X ppm be the measured PCB level in the composite. If $X \le (14.2/7) = 2.0$, then all 7 individual samples are rated clean. If $X > 14.2$, then at least one individual sample must be above the 10 ppm limit. If $2.0 < X \le 14.2$, no conclusion is possible based on analysis of the composite and the 7 samples must be analyzed individually to reach a decision. These results may be generalized to a composite of any arbitrary number of samples, subject to the limitations noted below.

The applicability of compositing is potentially limited by the size of the individual specimens and by the performance of the analytical method at low PCB levels. First, the individual specimens must be large enough so that the composite can be formed while leaving enough material for individual analyses if needed. For verification of PCB spill cleanup, adequacy of specimen sizes should not be a problem. The second limiting factor is the analytical method. Down to about 1 ppm, the performance of the stipulated analytical methods should not degrade markedly. Therefore, since the assumed permissible level is 10 ppm, no more than about 10 specimens should be composited at a time.

In compositing specimens, the location of the sampling points to be grouped should be taken into account. If a substantial residual area of contamination is present, then contaminated samples will be found close together. Thus, contiguous specimens should be composited, if feasible, in order to maximize the potential reduction in the number of analyses produced by the compositing strategy. Rather than describe a (very complicated) algorithm for choosing specimens to composite, we have graphically indicated some possible compositing strategies in Figures 8 Through 11. Based on the error probability calculations presented in Section 4 below, we recommend the compositing strategies indicated in Table 6. The recommended strategy for the 7-point design requires no explanation. The strategies for the 19- and 37-point cases are shown in Figures 9 and 11, respectively. The strategies shown in Figures 8 and 10 are used in Section 5 for comparison purposes. For details on the reduction in number of analyses expected to result (as compared to individual analyses), see the next Section, 5.

A 2 GROUP COMPOSITING PLAN FOR 7 SAMPLE POINTS

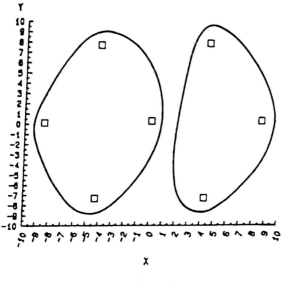

Figure 8

A 2 GROUP COMPOSITING PLAN FOR 19 SAMPLE POINTS

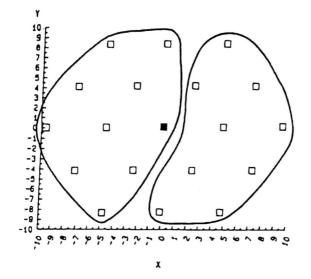

Figure 9

A 6 GROUP COMPOSITING PLAN FOR 19 SAMPLE POINTS

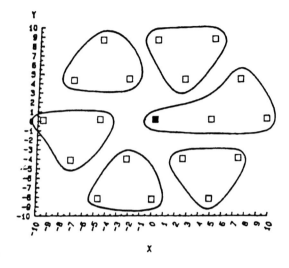

Figure 10. Location of sample points in a 19 sample point plan, with detail of a 2 group compositing design.

A 4 GROUP COMPOSITING PLAN FOR 37 SAMPLE POINTS

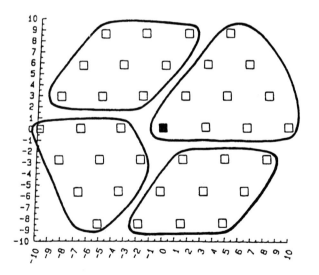

Figure 11. Location of sample points in 37 sample point plan, with detail of a 4 group compositing design.

5. *Calculations of Average Number of Analyses, and Error Probabilities*

Estimates of <u>expected number of analyses</u> and probabilities of <u>false positives</u> (incorrectly deciding the site is contaminated above the limit), and <u>false negatives</u> (failure to detect residual contamination) were obtained for various scenarios. The calculations were performed by Monte Carlo simulation using 5,000 trials for each combination of sample size, compositing strategy, level, and extent of residual contamination. The computations were based on the following assumptions:

Table 6. Recommended Compositing Strategies

No. of samples collected	Compositing strategy
7	One group of 7
19	One group of 10, one of 9
37	Three groups of 9, one of 10

a. Only soil samples are involved. In practice other types of samples will often be obtained and analyzed. Although the results of this section are not directly applicable to such cases, they do indicate in general terms the type of accuracy obtainable and the potential cost savings from compositing.

b. If the true PCB level in a sample is C, then the measured value is a normally distributed random variable with mean 0.8C and standard deviation $(0.3)(0.8C) = 0.24C$. Thus, it is assumed that the analytical method is 80% accurate, with 30% relative standard deviation.

c. The maximum allowable level in a single sample is 10 ppm. However, the <u>measured</u> level for a single sample must exceed 14.2 ppm for a finding of noncompliance. As previously discussed, this corresponds to a single-sample false- positive rate of 0.5%.

d. The residual contamination present is modeled as a randomly placed circle of variable radius and contamination level. The PCB level is assumed to be uniform within the randomly-placed circle and zero outside it.

e. Analysis of samples is terminated as soon as a positive result is obtained on a single analysis. If a composite does not give a definitive result (positive or negative), the individual specimens from which the composite was formed are analyzed in sequence before any other composite.

f. The compositing strategies used are shown in Figures 8 and 11.

The results of the computations are shown in Tables 7 through 20. Tables 7 through 12 show the performance of the compositing strategies recommended in Section 3. For each strategy, there is a pair of tables. The first table shows the probability of reporting a violation of a 10 ppm cleanup standard, for different levels of residual contamination and percent of cleanup area contaminated. When the contamination level is 10 ppm or less, the number in the table is the probability of a false positive, i.e., a false finding of noncompliance. These probabilities are all very low, as they should be. When the level is above 10 ppm, the number in the table is the probability that a violation will be detected by the sampling design. For levels close to 10 ppm, and for small percentages of cleanup area residually contaminated, the detection probability is low. When the level is high and the percent of area contaminated is large, however, detection probability approaches 100%. For small areas with high contamination, detection capability is modest. This is because there is only a small chance that the contaminated area will be sampled. Similarly, detection capability is also modest for large areas contaminated near the 10 ppm limit. The reason for this is that, even though a number of contaminated samples will be found in such cases, the analytical method is not likely to give positive identification of levels near the 10 ppm cutoff. This is the price paid for reducing the single-sample false positive rate to 0.5%.

The second table for each compositing strategy shows the expected (average) number of analyses needed to reach a decision. For a fixed percent of area contaminated, the smallest number of analyses is needed if the level of contamination is very high or very low. For intermediate levels, more analyses are needed. The largest number of analyses are required with a large area contaminated at close to 10 ppm. In such a situation, the levels of the composite(s) will mostly lie in the intermediate range for which no conclusion is possible based on analysis of the composite. Thus, individual analyses will almost always be required, so that the advantage of compositing is lost.

Tables 13 through 20 compare the recommended compositing strategies for the 7-point and 19-point designs to alternative compositing strategies for these designs, for 4 different contaminated percentages (1%, 9%, 25%, and 49%). The comparison is based on the expected number of analyses required. Overall detection capabilities are comparable for the different strategies. The tables show that the recommended strategies are best, except for larger areas contaminated close to the 10 ppm level.

Table 7. Probability of Declaring a Violation of a 10 ppm Cleanup Standard, for the 7 Point, 1 Composite Design[a]

Level of residual PCB contamination (ppm)		Percent of cleanup area with residual PCB contamination					
		1	4	9	16	25	49
Compliant	8	<0.001	<0.001	<0.001	<0.001	<0.001	<0.001
	10	<0.001	<0.001	<0.001	<0.001	0.002	0.007
Noncompliant	11	<0.001	<0.001	<0.001	<0.001	0.009	0.032
	12	<0.001	0.001	0.001	0.002	0.017	0.092
	13	0.001	0.005	0.005	0.009	0.045	0.184
	14	0.003	0.010	0.019	0.028	0.085	0.298
	15	0.006	0.016	0.039	0.065	0.134	0.396
	16	0.009	0.029	0.064	0.102	0.202	0.617
	18	0.019	0.074	0.137	0.218	0.344	0.655
	20	0.030	0.110	0.199	0.335	0.479	0.787
	25	0.048	0.186	0.342	0.554	0.736	0.905
	50	0.070	0.245	0.487	0.767	0.977	0.989
	75	0.071	0.245	0.496	0.787	0.992	0.995
	100	0.068	0.255	0.499	0.800	0.995	0.997
	150	0.070	0.246	0.481	0.796	0.998	0.999
	200	0.073	0.254	0.489	0.806	>0.999	>0.999
	300	0.069	0.257	0.494	0.792	>0.999	>0.999
	500	0.070	0.242	0.492	0.811	>0.999	>0.999

[a]Seven samples analyzed first as a composite, then individually if necessary to reach a decision.

Table 8. Expected Number of Analyses to Decide Compliance or Violation, for a 10 ppm Cleanup Standard, for the 7-Point, 1-Composite Design[a]

Level of residual PCB contamination (ppm)		Percent of cleanup area with residual PCB contamination					
		1	4	9	16	25	49
Compliant	4	1.00	1.00	1.00	1.00	1.00	1.11
	6	1.00	1.00	1.00	1.00	1.06	2.31
	8	1.00	1.00	1.00	1.00	1.44	3.96
	10	1.00	1.01	1.02	1.03	1.75	4.96
Noncompliant	11	1.01	1.04	1.05	1.11	2.01	5.31
	12	1.04	1.08	1.17	1.32	2.21	5.39
	13	1.04	1.18	1.40	1.59	2.56	5.35
	14	1.10	1.32	1.63	2.02	2.86	5.18
	15	1.13	1.45	1.85	2.35	3.22	4.90
	16	1.15	1.52	2.03	2.67	3.50	4.71
	18	1.19	1.69	2.41	3.18	3.95	4.36
	20	1.24	1.85	2.57	3.59	4.19	4.04
	25	1.26	1.98	2.85	3.84	4.47	3.61
	50	1.28	1.96	2.93	3.99	4.45	2.96
	75	1.28	1.94	2.93	3.98	4.23	2.26
	100	1.21	1.79	2.53	3.45	3.54	1.87
	150	1.09	1.28	1.52	1.86	1.89	1.30
	200	1.03	1.11	1.15	1.34	1.33	1.13
	300	1.01	1.01	1.04	1.09	1.06	1.03
	500	1.00	1.00	1.01	1.02	1.02	1.01

[a]Seven samples analyzed first as a composite, then individually if necessary to reach a decision.

Table 9. Probablility of Declaring a Violation of a 10 ppm Cleanup Standard, for the 19 Point, 2 Composite Design[a]

Level of residual PCB contamination (ppm)		Percent of cleanup area with residual PCB contamination					
		1	4	9	16	25	49
Compliant	8	<0.001	<0.001	<0.001	<0.001	<0.001	<0.001
	10	<0.001	<0.001	0.002	0.007	0.015	0.028
Noncompliant	11	<0.001	<0.001	0.007	0.034	0.058	0.017
	12	0.001	0.002	0.029	0.084	0.153	0.281
	13	0.003	0.007	0.062	0.179	0.304	0.497
	14	0.005	0.021	0.114	0.304	0.455	0.693
	15	0.012	0.052	0.178	0.407	0.606	0.832
	16	0.025	0.083	0.264	0.518	0.744	0.908
	18	0.046	0.167	0.421	0.698	0.883	0.978
	20	0.077	0.263	0.556	0.812	0.945	0.993
	25	0.125	0.461	0.784	0.923	0.990	0.999
	50	0.161	0.631	0.978	0.992	0.999	>0.999
	75	0.171	0.651	0.993	0.997	>0.999	>0.999
	100	0.168	0.642	0.994	0.999	>0.999	>0.999
	150	0.166	0.657	0.998	0.999	>0.999	>0.999
	200	0.175	0.648	0.999	0.999	>0.999	>0.999
	300	0.168	0.654	0.999	>0.999	>0.999	>0.999
	500	0.180	0.661	0.999	>0.999	>0.999	>0.999

[a]Nineteen samples analyzed first as two composites, then individually if necessary to reach a decision.

Table 10.   Expected Number of Analyses to Decide Compliance or Violation, for a 10 ppm Cleanup Standard, for the 19-Point, 2-Composite Design[a]

Level of residual PCB contamination (ppm)		Percent of cleanup area with residual PCB contamination					
		1	4	9	16	25	49
Compliant	4	2.00	2.00	2.00	2.18	3.30	7.49
	6	2.00	2.00	2.00	3.79	6.70	11.22
	8	2.00	2.00	3.01	6.15	9.20	13.18
	10	2.01	2.03	3.72	7.46	10.55	14.02
Noncompliant	11	2.03	2.14	4.07	7.90	10.74	13.81
	12	2.10	2.32	4.57	8.08	10.67	12.78
	13	2.21	2.74	4.84	7.94	9.95	11.00
	14	2.25	3.02	5.16	7.90	9.31	9.27
	15	2.37	3.40	5.50	7.65	8.42	7.80
	16	2.49	3.84	5.89	7.30	7.59	6.63
	18	2.60	4.36	6.11	6.57	6.29	5.02
	20	2.68	4.65	6.26	6.18	5.48	4.25
	25	2.82	5.02	6.20	5.45	4.57	3.36
	50	2.80	5.03	5.96	4.70	3.48	2.28
	75	2.80	5.05	5.69	3.68	2.63	1.84
	100	2.77	4.95	5.37	3.46	2.26	1.69
	150	2.53	3.94	3.99	2.59	1.80	1.46
	200	2.21	2.67	2.61	1.91	1.55	1.33
	300	1.99	1.89	1.70	1.50	1.34	1.19
	500	1.92	1.69	1.48	1.39	1.30	1.16

[a]Nineteen samples analyzed first as two composites, then individually if necessary to reach a decision.

Table 11. Probability of Declaring a Violation of a 10 ppm Cleanup Standard, for the 37 Point, 4 Composite Design[a]

Level of residual PCB contamination (ppm)	Percent of cleanup area with residual PCB contamination					
	1	4	9	16	25	49
Compliant 8	<0.001	<0.001	<0.001	<0.001	<0.001	<0.001
10	<0.001	0.002	0.010	0.022	0.031	0.060
Noncompliant 11	0.001	0.008	0.041	0.084	0.124	0.225
12	0.001	0.024	0.103	0.217	0.305	0.488
13	0.005	0.053	0.224	0.388	0.536	0.751
14	0.012	0.094	0.360	0.575	0.726	0.908
15	0.023	0.159	0.501	0.740	0.859	0.950
16	0.039	0.242	0.621	0.831	0.936	0.991
18	0.091	0.390	0.785	0.940	0.985	>0.999
20	0.147	0.542	0.884	0.981	0.996	>0.999
25	0.249	0.771	0.958	0.995	0.999	>0.999
50	0.340	0.976	0.997	0.999	0.999	>0.999
75	0.343	0.991	0.999	0.999	>0.999	>0.999
100	0.353	0.993	0.999	>0.999	>0.999	>0.999
150	0.339	0.997	>0.999	>0.999	>0.999	>0.999
200	0.357	0.996	>0.999	>0.999	>0.999	>0.999
300	0.344	0.997	>0.999	>0.999	>0.999	>0.999
500	0.348	0.999	>0.999	>0.999	>0.999	>0.999

[a]Thirty-seven samples analyzed first as four composites, then individually if necessary to reach a decision.

Table 12.   Expected Number of Analyses to Decide Compliance or Violation, for a 10 ppm Cleanup Standard, for the 37-Point, 4-Composite Design[a]

Level of residual PCB contamination (ppm)		Percent of cleanup area with residual PCB contamination					
		1	4	9	16	25	49
Compliant	4	4.00	4.01	4.41	6.72	9.85	15.69
	6	4.00	4.15	6.66	10.22	13.48	19.36
	8	4.00	4.77	9.01	12.76	15.98	22.08
	10	4.02	5.36	10.56	14.29	17.18	23.04
Noncompliant	11	4.07	5.69	10.87	14.29	16.93	21.28
	12	4.18	5.97	10.94	13.74	15.68	17.84
	13	4.35	6.28	10.56	12.74	13.44	13.54
	14	4.57	6.78	10.21	11.21	11.13	10.10
	15	4.73	7.04	9.60	9.71	9.33	7.78
	16	4.90	7.33	9.08	8.77	7.83	6.12
	18	5.09	7.59	8.02	7.05	6.16	4.71
	20	5.26	7.74	7.28	6.26	5.30	3.96
	25	5.34	7.55	6.53	5.28	4.37	3.08
	50	5.27	7.14	5.39	3.78	3.06	2.16
	75	5.23	6.84	4.31	3.04	2.55	1.90
	100	5.22	6.43	3.73	2.64	2.32	1.73
	150	4.55	4.89	3.02	2.37	2.07	1.57
	200	3.95	3.57	2.53	2.15	1.90	1.52
	300	3.59	2.67	2.28	2.04	1.81	1.44
	500	3.49	2.48	2.22	1.99	1.79	1.44

[a]Thirty-seven samples analyzed first as four composites, then individually if necessary to reach a decision.

Table 13.   Comparison of Expected Number of Analyses for Different Compositing Strategies for the 7-Point Design, When an Area 1% of the Size of the Cleanup Site Remains Contaminted.

Level of residual PCB contamination (ppm)		1 Composite	2 Composites	Individually
Compliant	4	1.00	2.00	7.00
	8	1.00	2.00	7.00
	10	1.00	2.00	7.00
Noncompliant	12	1.04	2.02	6.98
	14	1.10	2.05	6.96
	16	1.15	2.07	6.92
	20	1.24	2.10	6.88
	25	1.26	2.11	6.84
	50	1.28	2.09	6.80
	100	1.21	1.98	6.78
	200	1.03	1.96	6.80
	500	1.00	1.96	6.81

Table 14.   Comparison of Expected Number of Analyses for Different Compositing Strategies for the 7-Point Design, When an Area 9% of the Size of the Cleanup Site Remains Contaminted.

Level of residual PCB contamination (ppm)		1 Composite	2 Composites	Individually
Compliant	4	1.00	2.00	7.00
	8	1.00	2.00	7.00
	10	1.02	2.01	6.99
Noncompliant	12	1.17	2.09	6.91
	14	1.63	2.32	6.69
	16	2.03	2.50	6.49
	20	2.57	2.77	6.06
	25	2.85	2.79	5.65
	50	2.93	2.60	5.46
	100	2.53	1.85	5.46
	200	1.15	1.72	5.45
	500	1.01	1.17	5.45

Table 15.   Comparison of Expected Number of Analyses for Different Compositing Strategies for the 7-Point Design, When an Area 25% of the Size of the Cleanup Site Remains Contaminted.

Level of residual PCB contamination (ppm)		1 Composite	2 Composites	Individually
Compliant	4	1.00	2.00	7.00
	8	1.44	2.13	7.00
	10	1.71	2.24	6.98
Noncompliant	12	2.21	2.44	6.81
	14	2.86	2.84	6.29
	16	3.50	3.23	5.64
	20	4.19	3.54	4.68
	25	4.47	3.56	4.12
	50	4.45	2.97	3.58
	100	3.54	1.61	3.51
	200	1.33	1.38	3.50
	500	1.02	1.37	3.50

Table 16.   Comparison of Expected Number of Analyses for Different Compositing Strategies for the 7-Point Design, When an Area 49% of the Size of the Cleanup Site Remains Contaminted.

Level of residual PCB contamination (ppm)		1 Composite	2 Composites	Individually
Compliant	4	1.11	2.02	7.00
	8	3.96	2.99	7.00
	10	4.96	3.50	6.96
Noncompliant	12	5.39	3.81	6.61
	14	5.18	3.94	5.79
	16	4.71	3.86	4.82
	20	4.04	3.49	3.53
	25	3.61	3.03	2.87
	50	2.96	2.22	2.40
	100	1.87	1.36	2.40
	200	1.13	1.23	2.39
	500	1.01	1.20	2.39

Table 17.   Comparison of Expected Number of Analyses for Different Compositing Strategies for the 19-Point Design, When an Area 1% of the Size of the Cleanup Site Remains Contaminted.

Level of residual PCB contamination (ppm)		2 Composites	6 Composites	Individually
Compliant	4	2.00	6.00	19.00
	8	2.00	6.00	19.00
	10	2.01	6.00	19.00
Noncompliant	12	2.10	6.03	18.93
	14	2.25	6.07	18.74
	16	2.49	6.11	18.46
	20	2.68	6.07	18.06
	25	2.82	6.01	17.75
	50	2.80	5.80	17.49
	100	2.77	5.56	17.46
	200	2.21	5.53	17.46
	500	1.92	5.57	17.46

Table 18.   Comparison of Expected Number of Analyses for Different Compositing Strategies for the 19-Point Design, When an Area 9% of the Size of the Cleanup Site Remains Contaminted.

Level of residual PCB contamination (ppm)		2 Composites	6 Composites	Individually
Compliant	4	2.00	6.00	19.00
	8	3.01	6.19	19.00
	10	3.72	6.32	18.96
Noncompliant	12	4.57	6.54	18.40
	14	5.16	6.74	16.90
	16	5.89	6.83	14.86
	20	6.26	6.33	11.89
	25	6.20	5.74	10.22
	50	5.96	4.45	8.94
	100	5.37	3.34	8.64
	200	2.61	3.17	8.63
	500	1.48	3.17	8.62

Table 19.   Comparison of Expected Number of Analyses for Different Compositing Strategies for the 19-Point Design, When an Area 25% of the Size of the Cleanup Site Remains Contaminted.

Level of residual PCB contamination (ppm)		2 Composites	6 Composites	Individually
Compliant	4	3.30	6.07	19.00
	8	9.20	7.73	19.00
	10	10.55	8.44	18.83
Noncompliant	12	10.67	8.47	17.31
	14	9.31	7.67	13.72
	16	7.59	6.57	10.58
	20	5.48	5.09	6.25
	25	4.57	4.24	4.35
	50	3.48	3.22	3.34
	100	2.26	2.51	3.29
	200	1.55	2.41	3.26
	500	1.30	2.43	3.23

Table 20.   Comparison of Expected Number of Analyses for Different Compositing Strategies for the 19-Point Design, When an Area 49% of the Size of the Cleanup Site Remains Contaminted.

Level of residual PCB contamination (ppm)		2 Composites	6 Composites	Individually
Compliant	4	7.49	6.28	19.00
	8	13.18	9.85	19.00
	10	14.02	10.84	18.73
Noncompliant	12	12.78	10.10	16.15
	14	9.27	7.78	11.34
	16	6.63	5.87	7.14
	20	4.25	3.92	3.74
	25	3.36	3.23	2.61
	50	2.28	2.46	2.10
	100	1.69	1.85	2.06
	200	1.33	1.79	2.04
	500	1.16	1.78	2.02

The major conclusions that can be drawn from these results are as follows. First, the proposed cutoff on the measured PCB level for a finding of noncompliance for a single sample, 14.2 ppm, is successful in.controlling the overall false positive rate of the sampling scheme. For example, when an area half the size of the entire site remains contaminated just at the allowable limit of 10 ppm, the false positive rate is 1% for the 7-point design, 3% for the 19-point design, and 6% for the 37-point design. Note, that the overall false-positive rate is highest for contamination just at the allowable limit. Second, the detection capabilities of the design appear satisfactory, bearing in mind the difficulty of detecting randomly-located contamination by any sampling scheme without exhaustive sampling. As an example, the proposed 19-point design can detect 50 ppm contamination present in 9% of the cleanup area with 98% probability. Similarly, the 19-point design can detect 20 ppm contamination present in 25% of the area with 95% probability. Third, the proposed compositing strategies are quite effective in reducing the number of analyses needed to reach a decision in all cases except those involving large areas contaminated near the cutoff of 10 ppm. For example, for contaminated levels of 25 ppm or greater, the expected number of analyses to reach a decision never exceeds 5 for the 7-point design, or 7 for the 19-point design, or 8 for the 37-point design. Larger number of analyses are needed in cases of contamination close to the allowable limit of 10 ppm, up to 23 for the 37-point design when 49% of the area is contaminated at 10 ppm.

B. Sampling Techniques

The types of media to be sampled will include soil, water, vegetation and solid surfaces (concrete, asphalt, wood, etc.). General sampling methods are described below. Additional sampling guidance documents are available (Mason 1982, USWAG 1984).

1. Solids Sampling

When soil, sand, or sediment samples are to be taken, a surface scrape samples should be collected. Using a 10 cm x 10 cm (100 cm^2) template to mark the area to be sampled, the surface should be scraped to a depth of 1 cm with a stainless steel trowel or similar implement. This should yield at least 100 g soil. If more sample is required, expand the area but do not sample deeper. Use a disposable template or thoroughly clean the template between samples to prevent contamination of subsequent samples. The sample should be scraped directly into a precleaned glass bottle. If it is freeflowing, the sample should be thoroughly homogenized by tumbling. If not, successive subdivision in a stainless steel bowl should be used to create a representative subsample.

In some cases, such as sod, scrape samples may not be appropriate. For these cases, core samples, not more than 5 cm deep, should be taken using a soil coring device. These core samples should be well-homogenized in a stainless steel bowl by successive subdivision. A portion of each sample should then be removed, weighed and analyzed.

Samples should be stored in the dark at 4°C in precleaned glass bottles. If samples are to be analyzed quickly, the storage requirements may be relaxed as long as sample integrity is maintained. Before collection of verification samples, this equipment must be used to generate a field blank as described in Section IV.E.

2. Water Sampling

a. Surface Sampling

If PCBs dissolved in a hydrocarbon oil were spilled, they will most likely be dispersed on the surface. Therefore, a surface water collection technique should be used. Surface water samples should be collected by grab techniques. Where appropriate, the precleaned glass sample bottle may be dipped directly into the body of water at the designated sample collection point. A sample is collected from the water surface by gently lowering a precleaned sample bottle horizontally into the water until water begins to run into it. The bottle is then slowly turned upright keeFing the lip just under the surface so that the entire sample is collected from the surface.

b. Subsurface Sampling

If the PCBs were in an Askarel or other heavier-than-water matrix, the PCBs will sink. In these cases water near the bottom should be collected. To collect subsurface water, the bottle should be lowered to the specified depth with the cap on. The cap is then removed, the bottle allowed to fill, and the bottle brought to the surface.

c. Other Sampling Approaches

When the above approaches are not feasible, other dippers, tubes, siphons, pumps, etc., may be used to transfer the water to the sample bottle. The sampling system should be of stainless steel, Teflon, or other inert, impervious, and noncontaminating material. Before collection of samples, this equipment must be used to generate a field blank as described in Section IV.E.

d. Sample Preservation

The bottle is then lifted out of the water, capped with a PTFE- or foil-lined lid, identified with a sample number, and stored at approximately 4°C (USEPA 1984a) until analysis to retard bacterial growth. If samples are to be analyzed quickly, the storage requirements may be relaxed as long as sample integrity is maintained.

3. Surface Sampling

a. Wipe Samples

If the surface to be sampled is smooth and impervious (e.g., rain gutters, aluminum house siding), a wipe sample should indicate whether the cleanup has sufficiently removed the PCBs. These surfaces should be sampled by first applying an appropriate solvent (e.g., hexane) to a piece of 11 cm filterpaper (e.g., Whatman 40 ashless, Whatman "50" smear tabs, or equivalent) or gauze pad. This moistened filter paper or gauze pad is held with a pair of stainless steel forceps and used to thoroughly swab a 100-cm^2 area as measured by a sampling template.

Care must be taken to assure proper use of a sampling template. Different templates may be used for the variously shaped areas which must be sampled. A 100 cm^2 area may be a 10 cm x 10 cm square, a rectangle (e.g., 1 cm x 100 cm or 5 cm x 20 cm), or any other shape. The use of a template assists the sampler in the collection of a 100 cm^2 sample and in the selection of representative sampling sites. When a template is used it must be thoroughly cleaned between samples to prevent contamination of subsequent samples by the template.

The wipe samples should be stored in precleaned glass jars at 4°C. Before collection of verification samples, the selected filter paper or gauze pad and solvent should be used to generate a field blank as described in Section IV.E.

b. Sampling Porous Surfaces

Wipe sampling is inappropriate for surfaces which are porous and would absorb PCBs. These include wood and asphalt. Where possible, a discrete object (e.g., a paving brick) may be removed. Otherwise, chisels, drills, saws, etc., may be used to remove a sufficient sample for analysis. Samples less than 1 cm deep on the surface most likely to be contaminated with PCBs should be collected.

4. Vegetation Sampling

The sample design or visual inspection may indicate that samples of vegetation (such as leaves, bushes, and flowers) are required. In this case, samples may be taken with pruning shears, a saw, or other suitable tool and placed in a precleaned glass bottle.

C. Analytical Techniques

A number of analytical techniques have been used for analysis of PCBs in the types of samples which may be associated with PCB spills. Some of the candidate analytical methods are listed in Table 21. The analysis method(s) most appropriate for a given spill will depend upon a number of factors. These include sensitivity required, precision and accuracy required, potential interferents, ultimate use of the data, experience of the analyst, availability of laboratory equipment, and number of samples to be analyzed.

As shown in Table 21, many analytical methods are available. The general analytical techniques are discussed and then compared below.

1. Gas Chromatography (GC)

As can be seen in Table 21, analysis of PCBs by gas chromatography is frequently the method of. PCBs are chromatographed using either and may be detected using either packed or capillary columns and may be detected using either specific detectors or mass spectrometry. A comprehensive method for analysis of PCBs in transformer fluid and waste oils was developed by Bellar and Lichtenberg (1982). This method describes six different cleanup techniques, recommends three GC detectors, and suggests procedures for GC calibration and for measurement of precision and accuracy. This method also discusses several calculation methods.

a. Gas Chromatograph/Electron Capture Detection

Packed column gas chromatography with electron capture detection (GC/ECD) is generally the method of choice for analysis of spill site samples, transformer oils, and other similar matrices which must be analyzed for PCB content prior to disposal (Copland and Gohmann 1982). GC/ECD is very sensitive, highly selective against hydrocarbon background, and relatively inexpensive to operate. The technique is most appropriate when the PCB residue resembles an Aroclor® (Aroclor® is a registered trademark of Monsanto Company; the trademark designation is not used throughout this report) standard and other halogenated compounds do not interfere.

While it is considered a selective detector, ECD also detects non-PCB compounds such as halogenated pesticides, polychlorinated naphthalenes, chloroaromatics, phthalate and adipate esters, and other compounds. These compounds may be differentiated from PCBs only by chromatographic retention time. Elemental sulfur can interfere with PCB analysis in sediment and other samples which have been subjected to anaerobic degradation conditions. There are also common interferences which do not give discrete

Table 21. Standard Procedures of Analysis for PCBs

Procedure designation	Matrix	Extraction	Cleanup[c]	Determination method	Qualitative assessment	Quantitation method	LOD	QC discussed	Reference
D3534-80	Water	Hexane/ CH_2Cl_2	(Florisil) (Silica Gel)	PGC/ECD[d]	No	Total area or Webb-McCall	0.1 $\mu g/L$	No	ASTM, 1981a
608	Water	CH_2Cl_2	(Florisil) (S removal)	PGC/ECD	No	Area	0.04-0.15 $\mu g/L$	Yes	EPA, 1984a; Longbottom and Lichtenberg, 1982
625	Water	CH_2Cl_2	None	PGC/EIMS (CGC)	Yes	Area	30-36 $\mu g/L$	Yes	EPA, 1984b; Longbottom and Lichtenberg, 1982
304h	Water	Hexane/ CH_2Cl_2 (85/15)	Florisil/ silica gel (CH_3CN) (S removal)	PGC/ECD or HECD	Yes	Summed areas or Webb-McCall	NS	Yes	EPA, 1978
EPA (by-products)	Water	Several	Several	HRGC/EIMS	Yes	Ind. peaks	NS	Yes	Erickson et al., 1982, 1983d; EPA, 1984c
ANSI	Water	Hexane	(H_2SO_4) (Saponification) Alumina	PGC/ECD	No	Single peak or summed peaks	2 ppm	Yes	ANSI, 1974
Monsanto	Water	Hexane	Alumina	PGC/ECD	No	Individual or total peak heights	2 ppb	No	Moein, 1976

Table 21. Standard Procedures of Analysis for PCBs (CONTINUED)

Procedure designation	Matrix	Extraction	Cleanup[c]	Determination method	Qualitative assessment	Quantitation method	LOD	QC discussed	Reference
UK-DOE	Water	Hexane	Silica gel	PGC/ECD	No	NS	106 ng/L	No	UK-DOE, 1979; Devenish and Harling-Bowen, 1980
D3304-74	Air Water Soil, Sediment	DI Hexane H_2O/CH_3CN	(H_2SO_4) (Saponification) (Alumina)	PGC/ECD	No	Total area	NS	Yes	ASTM, 1981b
EPA (homolog)	Solids and liquids	Several	Several	HRGC/EIMS	Yes	Ind. peaks	NS	Yes	Erickson et al., 1985a
EPA 625-S	Sludge	CH_2Cl_2	Florisil, Silica gel, or GPC	HRGC/EIMS or PGC/EIMS	Yes	Area	NS	Yes	Haile and Lopez-Avila, 1984
EPA (Halocarbon)	Sludge	Hexane/ $CH_2Cl_2/$ acetone (83/15/2)	GPC S removal	PGC/ECD	Yes	Peak area or peak height	NS	Yes	Rodriguez et al., 1980
Priority Pollutant	Sludge	CH_2Cl_2 (base/ neutral and acid fractions)	GPC	PGC/EIMS	Yes	NS	NS	Yes	EPA, 1979c

Table 21. Standard Procedures of Analysis for PCBs (CONTINUED)

Procedure designation	Matrix	Extraction	Cleanup[c]	Determination method	Qualitative assessment	Quantitation method	LOD	QC discussed	Reference
8100	Sludge	CH_2Cl_2 (3 fractions)	GPC Silica gel	HRGC/EIMS or PGC/EIMS	Yes	NS	NS	Yes	Ballinger, 1978
8080	Solid Waste	CH_2Cl_2	(Florisil)	PGC/ECD	No	Area	1 µg/g	Yes	EPA, 1982e
8250	Solid Waste	CH_2Cl_2	None	PGC/EIMS	No	NS	1 µg/g	Yes	EPA, 1982e
8270	Solid Waste	$CHCl_2$	None	CGC/EIMS	No	NS	1 µg/g	Yes	EPA, 1982e
EPA (spills)	Unspecified	Hexane/acetone	(CH_3CN) (Florisil) (Silica gel) (Mercury)	PGC/ECD	No	Total area or Webb-McCall	NS	No	Beard and Schaum, 1978
EPA	Soil and Sediment	Acetone/Hexane	Florisil Silica gel (S removal)	PGC/ECD	No	Computer	NS	Yes	EPA, 1982d
Monsanto	Sediment	CH_3CN	Saponification H_2SO_4 Alumina	PGC/ECD	No	Individual or total peak heights	2 ppb	No	Moein, 1976
ANSI	Sediment, soil	CH_3CN	Saponification H_2SO_4 Alumina	PGC/ECD	No	Single peak or summed peaks	2 ppm	Yes	ANSI, 1974

Table 21. Standard Procedures of Analysis for PCBs (CONTINUED)

Procedure designation	Matrix	Extraction	Cleanup[c]	Determination method	Qualitative assessment	Quantitation method	LOD	QC discussed	Reference
EPA (by-products)	Air collected on Florisil or XAD-2	Hexane	(H₂SO₄) (Florisil)	HRGC/EIMS	Yes	Ind. peaks	NS	Yes	Erickson et al., 1982, 1983d; Erickson, 1984b
EPA (ambient air)	Air near hazardous waste sites collected on PUF	Hexane/ ether	Alumina	PGC/ECD	No	Total area or peak height	10-50 ng/m³	No	Lewis, 1982
EPA (stack)	Incinerator emissions and ambient air collected on Florisil	Hexane	(H₂SO₄)	Perchlorina-tion PGC/ECD	No	Area	10 ng	No	Haile and Baladi, 1977; Beard and Schaum, 1978
EPA	Combustion sources collected on Florisil	Pentane or CH₂Cl₂	(Florisil/ silica gel)	PGC/MS	Yes	Area/homolog	0.1 ng/inj	No	Levins et al., 1979
EPA (incin-erators)	Stack gas	Pentane/ methanol		PGC/MS	Yes	Single peak	NS	Yes	Beard and Schaum, 1978

Table 21. Standard Procedures of Analysis for PCBs (CONTINUED)

Procedure designation	Matrix	Extraction	Cleanup[c]	Determination method	Qualitative assessment	Quantitation method	LOD	QC discussed	Reference
ANSI	Air (toluene impinger)		(H_2SO_4) (Spanofication) (Alumina)	PGC/ECD	No	Single peak	2 ppb	Yes	ANSI, 1974
NIOSH (P&CAM 244)	Air collected on Florisil	Hexane	None	PGC/ECD	No	Peak height or area from standard curve or Webb-McCall	0.01 mg/m^3	No	NIOSH, 1977a
NIOSH (P&CAM 253)	Air collected on Florisil	Hexane	None	PGC/ECD Perchlorination	No	Peak height or area from standard curve	0.01 mg/m^3	No	NIOSH, 1977b,c
EPA (gas)	Natural gas sampled with Florisil	Hexane	H_2SO_4	PGC/ECD	No	Total area, peak height or Webb-McCall (Perchlorination)	0.1-2 µg/m^3	No	Harris et al., 1981
EPA [5,A,(3)]	Blood	Hexane	(Florisil)	PGC/ECD	No	NS	NS	No	Watts, 1980
EPA [5,A,(1)]	Adipose	Pet. ether/CH$_3$CN	Florisil	PGC/ECD	No	NS	NS	Yes	Watts, 1980
EPA (9,D)	Adipose	Pet. ether/CH$_3$CN	Saponification Florisil	TLC	No	Semiquant.	10 ppm	No	Watts, 1980

Table 21. Standard Procedures of Analysis for PCBs (CONTINED)

Procedure designation	Matrix	Extraction	Cleanup[c]	Determination method	Qualitative assessment	Quantitation method	LOD	QC discussed	Reference
EPA (9,B)	Milk	Acetone/hexane	CH_3CN Florisil Silica acid	PGC/ECD	Yes	Ind. peaks	50 ppb	Yes	Watts, 1980 Sherma, 1981
AOAC (29)	Food	CH_3CN/Pet. ether	Florisil MgO/ Celite Saponification	PGC/ECD	No	Total area or Ind. peaks	NS[a]	No	AOAC, 1980a
Japan	Food	Pet. ether/ CH_3CN	Silica gel Saponification (Florisil)	PGC/ECD	Yes	Summed areas perchlorination	NS	No	Tanabe, 1976
PAM	Food	Pet. ether/ CH_3CN	Silicic acid (Saponification) Oxidation (Florisil)	PGC/ECD (PGC/HECD) (NP-TLC) (RP-TLC)	No	Area	NS	No	FDA, 1977
AOAC (29)	Paper and paperboard	Saponification	Florisil MgO/ Celite Saponification	PGC/ECD	No	Total area or Ind. peaks	NS[e]	No	AOAC, 1980b
D3303-74	Capacitor skarels	DI[b]	None	SCOT HRGC/FID	No	Total area	2.8×10^{-8} mol/L	No	ASTM, 1980a

Table 21. Standard Procedures of Analysis for PCBs (CONTINUED)

Procedure designation	Matrix	Extraction	Cleanup[c]	Determination method	Qualitative assessment	Quantitation method	LOD	QC discussed	Reference
D4059-83	Mineral oil	Dilute with hexane or Isooctane	Florisil slurry (H_2SO_4) (Florisil column)	PGC/ECD (PGC/HECD)	Yes	Ind. peaks or Webb-McCall	50 ppm	No	ASTM, 1983
EPA (oil)	Transformer fluids or waste oils	DI	(H_2SO_4) (Florisil) (Alumina) (Silica gel) (GPC), (CH_3CN)	PGC/HECD or /ECD or /EIMS (HRGC)	No	Total area or Webb-McCall	1 mg/kg	Yes	EPA, 1981 Bellar and Lichtenberg, 1981
EPA (by-products)	Products or wastes	Several	Several	HRGC/EIMS	Yes	Ind. peaks	NS	Yes	Erickson et al., 1982, 1983d; Erickson, 1984a
DCMA	3 pigment types	A. Hexane/ H_2SO_4 B. CH_2Cl_2	None Florisil	PGC/ECD	No	10 isomers	~ 1 ppm/homolog	Yes	DCMA, 1982
DOW	Clorinated benzenes	DI	None	PGC/EIMS	Yes	Total peak height/homolog	NS	Yes	Dow, 1981

Table 21. Standard Procedures of Analysis for PCBs (CONTINUED)

Procedure designation	Matrix	Extraction	Cleanup[c]	Determination method	Qualitative assessment	Quantitation method	LOD	QC discussed	Reference
EPA (isomer groups)	Unspecified	Not addressed	Not addressed	HRGC/EIMS	Yes	Ind. peaks	NS	Yes	EPA, 1984d

Source M. D. Erickson, The Analytical Chemistry of PCBs, Butterworths, Boston, MA, 1985, in press.

[a]No specific details.
[b]Direct injection or dilute and inject.
[c]Techniques in parentheses are described as optional in the procedure.
[d]Or PGC with microcoulometric or electrolytic conductivity.

peaks. An example of a nonspecific interference is mineral oil (ASTM 1983). Mineral oil, a complex mixture of hydrocarbons, can cause a general suppression of ECD response. Mineral oils from transformers often contain PCBs as a result of cross-contamination of transformer oils.

A major disadvantage of ECD is the range of response factors which different PCB congeners exhibit. Zitko et al. (1971) and Hattori et al. (1981) published response factors ranges of about 540 and 9000, respectively, Boe and Egaas (1979), Onsuka et al. (1983) and Singeretal. (1983) have also published ECD response factors. The range of response factors seriously inhibits reliable quantitation of individual PCB congeners or non-Aroclor PCBs unless the composition of the sample and standard are the same.

When PCBs are analyzed by packed column gas chromatography, the PCBs are usually quantitated by total areas or individual peaks. In the total areas method, the areas of all peaks in a retention window are summed and this total compared with the corresponding response of an Aroclor standard. With the individual peak quantitation method, response factors are calculated for each peak in the packed column chromatogram. The most prominent individual peak quantitation method was originated by Webb and McCall (1973). These results may be reported as an Aroclor concentration or as total PCB. Packed column GC techniques are generally useful for quantitation of samples which resemble pure Aroclors but are prone to errors from interfering compounds or from PCB mixtures that do not resemble pure Aroclors (Albro 1979). For this reason analysts have been using capillary gas chromatography for the analysis of PCBs. Capillary gas chromatography offers the analyst the ability to separate most of the individual PCB isomers. Bush et al. (1982) has proposed a method of obtaining "total PCB" values byintegration of all PCB peaks, using response factors generated from an Aroclor mixture. Zell and Ballschmiter (1980) have developed a simplified approach where only a selected few "diagnostic peaks" are quantitated. In a similar approach Tuinstra et al. (1983) have quantitated six specific, diagnostic congeners which appear to be useful for regulatory cutoff analyses.

   b. GC/Hall Electrolytic Conductivity Detector

Electrolytic conductivity detectors have also been used with packed column gas chromatography to selectively detect PCBs (Webb and McCall 1973, Sawyer 1978). The Hall electrolytic conductivity detector (HECD) measures the change in conductivity of a solution containing HCl or HBr which is formed by pyrolysis of halogenated organic GC effluents. The HECD exhibits 10^5-10^6 selectivity for halogenated compounds over other compounds. It also gives a linear response over at least a 10^3 range. HECD and ECD were compared for their use in detecting PCBs in waste oil, hydraulic fluid, capacitor fluid, and transformer oil (Sonchik et al. 1984). They found both detectors acceptable, but noted that the HECD gave higher results with less precision than the ECD. The method detection limits ranged from 3-12 ppm for HECD and 2-4 ppm for ECD. Greater than 100% recovery of spikes analyzed by HECD indicated a nonspecific response to non-PCB components, since extraneous peaks were not observed. Another comparison of HECD and ECD for the analysis of PCBs in oils at the 30-500 ppm levels found that the type of detector made no significant difference in the results (Levine et al. 1983). The authors noted that they had expected higher accuracy from the more specific HECD. They postulated that the cleanup procedures (Florisil, alumina, and sulfuric acid) all had effectively removed the non-PCB species which would have caused interferences in the ECD and reduced its accuracy.

   c. GC/Mass Spectrometry

Highly specific identification of PCBs is performed by GC with mass spectrometric (GC/MS) detection. High resolution gas chromatography is generally used with mass spectrometry, so individual PCB isomers may be separated and identified. A GC/MS produces a chromatogram consisting of data points at about 1 second intervals, which are actually full mass spectra. The data are stored by a computer and may be retrieved in a variety of ways. The data file contains information on the amount of compound (signal intensity), molecular weight (parent ion), and chemical composition (fragmentation patterns and isotopic clusters).

GC/MS is particularly suited to detection of PCBs because of its intense molecular ion and the characteristic chlorine cluster. Chlorine has two naturally occurring isotopes ^{35}Cl and ^{37}Cl, which occur in a ratio of 100:33. Thus, a molecule with one chlorine atom will have a parent ion, M, and an M+2 peak at 33% relative intensity. With two chlorine atoms, M+2 has an intensity of 66% and M+4, 11%.

Because of its expense, complexity of data, and lack of sensitivity, GC/MS has not been used as extensively as other GC methods (particularly GC/ECD), despite its inherently higher information content. As the above factors have been improved, GC/MS has become much more popular for analysis of PCBs, and will probably continue to increase in importance. Several factors including the introduction of routine instruments without costly accessories, decreasing data system costs, and mass-marketing, have combined to keep the costs of GC/MS down while prices of other instruments have risen steadily. With larger data systems and more versatile and "user-friendly" software, the large amount of data is more easily handled. However, data reduction of a GC/MS chromatogram still requires substantially more time than for a GC/ECD chromatogram. In addition, the sensitivity of GC/MS has improved.

d. Field-Portable Gas Chromatography Instrumentation

Gas chromatography may be used for analysis of samples in the field. Gas chromatography is a well-established laboratory technique, and portable instruments with electron capture detectors are available (Spittler 1983, Colby et al. 1983, Picker and Colby 1984). A field-portable GC/ECD was used to obtain rapid measurements of PCBs in sediment and soil (Spittler 1983). The sample preparation consisted of a single solvect extraction. The PCBs were eluted from the GC within 9 min. In a 6-h period, 40 soils and 10 QC samples were analyzed, with concentrations ranging from 0.2 to 24,000 ppm. The use of field analysis permits real-time decisions in a cleanup operation and reduces the need for either return visits to a site.

Mobile mass spectrometers are also available. An atmospheric pressure chemical ionization mass spectrometer, marketed by SCIEX, has been mounted in a van and used for in situ analyses of soil and clay (Lovett et al. 1983). The instrument has apparently been used for field determination of PCBs in a variety of emergency response situations, including hazardous waste site cleanups. Other, more conventional mass spectrometers, should also be amenable to use in the field.

2. *Thin-Layer Chromatography (TLC)*

Thin-layer chromatography is a well-established analytical technique which has been used for the determination of PCBs for many years. Since the publication of a TLC method for PCBs by Mulhern (Mulhern 1968, Mulhern et al. 1971); several researchers have used TLC to measure PCBs in various matrices. Methods have been reported by Willis and Addison (1972) for the analysis of Aroclor mixtures, by Piechalak (1984) for the analysis of soils, and by Stahr (1984) for the analysis of PCB containing oils. Even with a densitometer to measure the intensity of the spots, TLC is not generally considered quantitative. Order-of-magnitude estimates of the concentration are certainly obtainable, but the precision and accuracy probably do not approach that of the gas chromatographic methods.

A spill site sample extract will probably need to be cleaned up before TLC analysis. Levine et al. (1983) have published a comparison of various cleanup procedures. Stahr (1984) has compared the Levine sulfuric acid cleanup to a SepPak® C_{18} cleanup method.

Different TLC techniques have been used to improve the sensitivity and selectivity of the method. Several researchers have reported that the use of reverse-phase TLC (C_{18} bonded phase) achieves a better separation of PCBs from interferences (DeVos and Peet 1971, DeVos 1972, Stalling and Huckins 1973, Brinkman et al. 1976). Koch (1979) has reported an order of magnitude improvement in the PCe limit of detection through use of circular TLC. The two most common methods of visualization are fluorescence (Kan et al. 1973, Ueta et al. 1974) and reaction with $AgNO_3$ followed by UV irradiation (DeVos and Peet 1971, DeVos 1972, Kawabata 1974, Stahr 1984).

No direct comparison of the performance of TLC with other techniques for analysis of samples from spill sites has been made. Two studies (Bush et al. 1975, Collins et al. 1972) compared TLC and GC/ECD. In both studies, the PCB values obtained were comparable. However, the study by Bush et al. indicated that the TLC results were generally lower than GC/ECD.

3. Total Organic Halide Analyses

Total organic halide analysis can be used to estimate PCB concentrations for guiding field work, but is not appropriate for verification or enforcement analyses. A total organic halide analysis indicates the presence of chlorine and sometimes the other halogens. Many of the techniques also detect inorganic chlorides such as sodium chloride. The reduction of organochlorine to free chloride ion with metallic sodium can be used for PCB analysis. The free chloride ions can be then detected calorimetrically (Chlor-N-Oil®) or by a chloride ion-specific electrode (McGraw-Edison). The performance of these kits has not been tested with any matrix other than mineral oil. X-ray fluorescence (XRF) has also been studied as a PCB screening technique (McQuade 1982, Schwalb and Marquez 1982).

D. Selection of Appropriate Methods

1. Criteria for Selection

The primary criterion for an enforcement method is that the data be highly reliable (i.e., they are legally defensible). This does not necessarily imply that the most exotic, state-of-the-art methods be employed; rather that the methods have a sgund scientific basis and validation data to support their use. Many other criteria also enter into selection of a method, including accuracy, precision, reproducibility, comparability, consistency across matrices, availability, and cost.

For PCB spills, it is assumed that the spills will be relatively fresh and therefore that PCB mixtures will generally resemble those in commercial products (i.e., Aroclon®). It is further assumed that, for most of the matrices likely to be encountered, the levels of interferences will be relatively low.

2. Selection of Instrumental Techniques

Based upon the above criteria and assumptions, either GC/ECD or GC/MS should provide suitable data. Since GC/ECD is included in more standard methods and since the technique is more widely used, it appears to be the technique of choice. The primary methods recommended below are all based on GC/ECD instrumental analysis. Some of the secondary and confirmatory techniques are based on GC/EIMS.

3. Selection of Methods

Ideally, a standard method would be available for each matrix likely to be encountered in a PCB spill. The matrices of concern include solids (soil, sand, sediment, bricks, asphalt, wood, etc.), water, oil, surface wipes, and vegetation. The methods for these matrices are summarized in Table 22 and discussed in detail below. A primary recommended method is given and should be used in most spill instances. The secondary method may be useful for confirmatory analyses, or where the situation (e.g., high level of interferences) indicates that the primary method is not applicable. The methods used must be documented or referenced.

Table 22. Summary of Recommended Analytical Methods

| Matrix | Primary method (GC/ECD) | | Secondary method | | |
	Designation	Reference	Designation	GC detector	Reference
Solids	8080	USEPA 1982e	8250, 8270	MS	USEPA 1982e
Water	608	USEPA 1984a	625	MS	USEPA 1984b
Oil	"oil"	USEPA 1981a; Bellar and Lichtenberg, 1981	"oil"	MS	USEPA 1981a; Bellar and Lichtenberg, 1981
Surface wipes	Hexane extraction/ 608	None	Hexane extraction/625	MS	None
Vegetation	AOAC (29)	AOAC 1980a	None	None	None

a. Solids (Soil, Sand, Sediment, Bricks, Asphalt, Wood, Etc.)

EPA Method 8080 from SW-846 (USEPA 1982e) is the primary recommended method. The secondary methods, Method 8250 and Method 8270, are GC/NS analogs. Method 8080 entails an acetoGe/hexane (1:1) extraction, a Florisil column chromatographic cleanup, and a GC/ECD instrumental determination. A total area quantitation versus Aroclor standards is specified. No qualitative criteria are supplied. A detection limit of 1 pg/g is prescribed. No validation data are available.

Bulk samples (bricks, asphalt, wood, etc.) should be readily extractable using a Soxhlet extractor according to EPA Method 8080 (USEPA 1982e). The sample must be crushed and subsampled to ensure proper solvent contact.

b. Water

EPA Method 608 (USEPA 1984e) is recommended as the primary method. This is one of the "priority pollutant" methods and involves extraction of water samples with dichloromethane. An optional Florisil column chromatographic cleanup and also an optional sulfur removal are given. Samples are analyzed by GC/ECD and quantitated against the total area of Aroclor standards. No qualitative criteria are given. This method has been extensively validated and complex recovery and precision equations are given in the method for seven Aroclor mixtures. The average recovery is about 86% and average overall precision about $\pm$ 26%. The average recovery and precision for the more common Aroclors (1242, 1254, and 1260) are about 78% and $\pm$ 26%, respectively. Detection limits are not given in the current version (USEPA 1984a), although they were listed as between 0.04 and 0.15 μg/L for the seven Aroclor mixtures listed as priority pollutants in the method validation study (Millar et al. 1984).

c. Oils

Spilled oil samples should be analyzed according to an EPA method (Bellar and Lichtenberg 1981). The method is written for transformer fluids and waste oils, but should also be applicable to other similar oils such as capacitor fluids. In this method, samples are diluted by an appropriate factor (e.g., 1:1000). Six optional cleanup techniques are given. The sample may be analyzed by GC/ECD as the primary method. Secondary instrumental choices, also presented in the method, are GC/HECD, GC/MS, and capillary GC/MS. PCBs are quantitated by either total areas or the Webb-McCall (1973) method. No qualitative criteria are given. QC criteria are given. A detection limit of 1 mg/kg is stated, although it is highly dependent on the amount of dilution required. An interlaboratory validation study (Sonchik and Ronan 1984) indicated 81 to 126% recoveries for different PCB mixtures, with an average of 97% for Aroclors 1242, 1254, and 1260, as measured by ECD. The overall method precision ranged from $\pm$ 11 to $\pm$ 55%, with an average of $\pm$ 12% for Aroclors 1242, 1254, and 1260. The method validation statistics were presented in more detail as regression equations.

d. Surface Wipes

No standard method is available for analysis of PCBs collected on surface wipes. However, since this matrix should be relatively clean and easily extractable, a simple hexane extraction should be sufficient. Samples should be analyzed according to EPA Method 608 (USEPA 1984a), except for Section

10.1 through 10.3. In lieu of these sections, the sample should be extracted three times with 25 to 50 mL of hexane. The sample can be extracted by shaking for at least 1 min per extraction in the wide-mouthed jar used for sample storage. Note that the rinses should be with hexane so that solvent exchange from methylene chloride to hexane (Section 10.7) is not necessary.

e. Vegetation

The AOAC (1980a) procedure for food is recommended for analysis of vegetation (leaves, vegetables, etc.). This method involves extraction of a macerated sample with acetonitrile. The acetonitrile is diluted with water and the PCBs extracted into petroleum ether. The concentrated extract is cleaned up by Florisil column chromatography by elution with a mixture of ethyl ether and petroleum.ether. The sample is analyzed by GC/ECD with quantitation by total areas or individual peak heights as compared to Aroclor standards. No qualitative criteria are given. Validation studies with chicken fat and fish (Sawyer 1973) are not relevant to the types of matrices to be encountered in PCB spills.

4. Implementation of Methods

Each laboratory is responsible for generating reliable data. The first step is preparation of an in-house protocol. This detailed "cookbook" is based on methods cited above, but specifies which options must be followed and provides more detail in the conduct of the techniques. It is essential that a written protocol be prepared for auditing purposes.

Each laboratory is responsible for generating validation data to demonstrate the performance of the method in the laboratory. This can be done before processing of samples; however, it is often impractical. Validation of method performance (replicates, spikes, QC samples, etc.) while analyzing field samples is acceptable.

Changes in the above methods are acceptable, provided the changes are documented and also provided that they do not affect performance. Some minor changes (e.g., substitution of hexane for petroleum ether) do not generally require validation. More significant changes (e.g., substitution of a HECD for ECD) will require documentation of equivalent performance.

E. Quality Assurance

Quality assurance must be applied throughout the entire monitoring program including the sample planning and collection phase, the laboratory analysis phase, and the data processing and interpretation phase.

Each participating EPA or EPA contract laboratory must develop a quality assurance plan (QAP) according to EPA guidelines (USEPA 1980). Additional guidance is also available (USEPA 1983). The quality assurance plan must be submitted to the regional QA officer or other appropriate QA official for approval prior to analysis of samples.

1. Quality Assurance Plan

The elements of a QAP (U.S. EPA, 1980) include:

- Title page
- Table of contents
- Project description
- Project organization and responsibility
- QA objectives for measurement data in terms of precision, accuracy, completeness, representativeness, and comparability
- Sampling procedures
- Sample tracking and traceability
- Calibration procedures and frequency
- Analytical procedures
- Data reduction, validation and reporting
- Internal quality control checks
- Performance and system audits
- Preventive maintenance
- Specific routine procedures used to assess data precision, accuracy and completeness
- Corrective action
- Quality assurance reports to management

2. Quality Control

Each laboratory that uses this method must operate a formal quality control (QC) program. The minimum requirements of this program consist of an initial and continuing demonstration of acceptable laboratory performance by the analysis of check samples, spiked blanks, and field blanks. The laboratory must maintain performance records which define the quality of data that are generated.

The exact quality control measures will depend on the laboratory, type and number of samples, and client requirements. The QC measures should be stipulated in the QA Plan. The QC measures discussed below are given for example only. Laboratories must decide on which of the measures below, or additional measures, will be required for each situation.

a. Protocols

Virtually all of the available PCB methods contain numerous options and general instructions. Effective implementation by a laboratory requires the preparation of a detailed analysis protocol which may be followed unambiguously in the laboratory. This document should contain working instructions for all steps of the analysis. This document also forms the basis for conducting an audit.

b. Certification and Performance Checks

Prior to the analysis of samples, the laboratory must define its routine performance. At a minimum, this must include demonstration of acceptable response factor precision with at least three replicate analyses of a calibration solution; and analysis of a blind QC check sample (e.g., the response factor calibration solution at unknown concentration submitted by an independent QA officer). Acceptable criteria for the precision and the accuracy of the QC check sample analysis must be presented in the QA plan.

Ongoing performance checks should include periodic repetition of the initial demonstration or more elaborate measures. More elaborate measures may include control charts and analysis of QA check samples containing unknown PCBs, and possibly with matrix interferences.

c. Procedural QC

The various steps of the analytical procedure should have quality control measures. These include, but are not limited to, the following:

Instrumental Performance: Instrumental performance criteria and a system for routinely monitoring the performance should be set out in the QA Plan. Corrective action for when performance does not meet the criteria should also be stipulated.

Qualitative Identification: Any questionable results should be confirmed by a second analytical method. A least 10% of the identifications, as well as any questionable results, should be confirmed by a second analyst.

Quantitation: At least 10% of all calculations must be checked. The results should be manually checked after any changes in computer quantitation routines.

d. Sample QC

Each sample and each sample set must have QC measures applied to it to establish the data quality for each analysis result. The following should be considered when preparing the QA plan:

Field Blanks: Field blanks are analyzed to demonstrate that the sample collection equipment has not been contaminated. A field blank may be generated by using the sampling equipment to collect a blank sample (e.g., using the water sampling equipment to sample laboratory reagent grade water) or by extracting the sampling equipment (e.g., extracting a sheet of filter paper from the lot used to collect wipe samples or rinsing the soil sampling apparatus into the sample jar). A field blank must be collected and analyzed for each type of sample collected.

Laboratory Reagent Blanks: These blanks are generated in the laboratory and are analyzed to assess contamination of glassware, reagents, etc., in the laboratory. Generally, a reagent blank is processed through the entire analysis process. Although in special circumstances, additional reagent blanks may be generated which are processed through only part of the procedure to isolate sources of contamination. At least one laboratory reagent blank must be generated and analyzed for each type of sample analyzed.

Check Samples: These samples contain known concentrations of PCBs in the sample matrix. They are analyzed along with field samples to demonstrate the method performance. The PCB concentrations may be known to the analyst.

Blind Check Samples: These samples are the same as the check samples discussed above, except the PCB concentration is not known to the analyst.

Replicate Samples: One sample from each batch of 20 or fewer will be analyzed in triplicate. The sample is divided into three replicate subsamples and all these subsamples carried through the analytical procedure, blind to the analyst. The results of these analyses must be comparable within the limits required for spiked samples.

Spiked Samples: The sensitivity and reproducibility must be demonstrated for any method used to report verification data. This can be done by analyzing spiked blanks near the required detection limit. To demonstrate the ability of the method to reproducibly detect the spiked sample, one or more spiked samples should be analyzed in at least triplicate for each group of 20 or fewer samples within each sample type collected. Samples will be spiked with a PCB mixture similar to that spilled (e.g., Aroclor 1260). Example concentrations are:

Matrix	Spike Level
Soil, etc.	10 μg/g (10 ppm)
Water	100 μg/L (100 ppb))
Wipes	100 μg/wipe (100 μg/100 cm^2)

Quantitative techniques must detect the spike level within $\pm 30\%$ for all spiked samples.

e. Sample Custody

As part of the Quality Assurance Plan, the chain-of-custody protocol must be described. A chain-of-custody provides defensible proof of the sample and data integrity. The less rigorous sample traceability documentation merely provides a record of when operations were performed and by whom. Sample traceability is not acceptable for enforcement activities.

Chain-of-custody is required for analyses which may result in legal proceedings and where the data may be subject to legal scrutiny. Chain-of-custody provides conclusive written proof that samples are taken, transferred, prepared, and analyzed in an unbroken line as a means to maintain sample integrity. A sample is in custody if:

- It is in the possession of an authorized individual;
- It is in the field of vision of an authorized individual;
- It is in a designated secure area; or
- It has been placed in a locked container by an authorized individual.

A typical chain-of-custody protocol contains the following elements:

1. Unique sample identification numbers.

2. Records of sample container preparation and integrity prior to sampling.

3. Records of the sample collection such as:

   - Specific location of sampling.
   - Date of collection.
   - Exact time of collection.
   - Type of sample taken (e.g., air, water, soil).
   - Initialing each entry.
   - Entering pertinent information on chain-of custody record.
   - Maintaining the samples in one's possession or under lock and key.
   - Transporting or shipping the samples to the analysis laboratory.
   - Filling out the chain-of-custody records.
   - The chain-of-custody records must accompany the samples.

4. Unbroken custody during shipping. Complete shipping records must be retained; samples must be shipped in locked or sealed (evidence tape) containers.

5. Laboratory chain-of-custody procedures consist of:

   - Receiving the samples.

- Checking each sample for tampering.
- Checking each sample against the chain-of-custody records.
- Checking each sample and noting its condition.
- Assigning a sample custodian who will be responsible for maintaining chain-of-custody.
- Maintaining the sign-offs for every transfer of each sample on the chain-of-custody record.
- Ensuring that all manipulations of the sample are duly recorded in a laboratory notebook along with sample number and date. These manipulations will be verified by the program manager or a designee.

F. Documentation and Records

Each laboratory is responsible for maintaining complete records of the analysis. A detailed documentation plan should be prepared as part of the QAP. Laboratory notebooks should be used for handwritten records. Digital or other GC/MS data must be archived on magnetic tape, disk, or a similar device. Hard copy printouts may also be kept if desired. Hard copy analog data from strip chart recorders must be archived. QA records should also be retained.

The documentation must completely describe how the analysis was performed. Any variances from a standard protocol must be noted and fully described. Where a procedure lists options (e.g., sample cleanup), the option used and specifics (solvent volumes, digestion times, etc.) must be stated.

The remaining samples and extracts should be archived for at least 2 months or until the analysis report is approved by the client organization (whichever is longer) and then disposed unless other arrangements are made. The magnetic disks or tapes, hard copy chromatograms, hard copy spectra, quantitation reports, work sheets, etc., must be archived for at least 3 years. All calculations used to determine final concentrations must be documented. An example of each type of calculation should be submitted with each verification spot.

G. Reporting Results

Results of analysis will normally be reported as follows:

Matrix	Reporting Units
Soil, etc.	μg PCB/g of sample (ppm)
Water	mg PCB/L of sample (ppm)
Surfaces (wipes)	μg PCB/wipe (μg PCB/100 cm^2)

In some cases, the results are to be reported by homolog. In this case, 11 values are reported per sample: one each for the 10 homologs and one for the total. Some TSCA analyses require reporting the results in terms of resolvable gas chromatographic peak (U.S. EPA, 1982c, 1984e). In these cases, the number of results reported equals the number of peaks observed on the chromatogram. These analyses are generally associated with a regulatory cutoff (e.g., 2 μg/g per resolvable chromatographic peak (U.S. EPA, 1982c, 1984). In these cases it may be sufficient, depending on the client organization's request, to report only those peaks which are above the regulatory cutoff.

Even if an Aroclor is used as the quantitation standard, the results are never to be reported as "μg Aroclor®/g sample." TSCA regulates all PCBs, not merely a specific commercial mixture.

V. References

Albro PW. 1979. Problems in analytic methodology: sample handling, extraction, and cleanup. Ann NY Acad Sci 320:19-27.

American National Standards Institute, Inc. 1974. American national standard guidelines for handling and disposal of capicator- and transformer-grade askarels containing polychlorinated biphenyls. ANSI C107.1-1974. New York, NY.

American Society for Testing and Materials. 1980. Standard method for rapid gas chromatographic estimation of high boiling homologues of chlorinated biphenyls for capacitor askarels. ANSI/ASTM D 3303-74 (Reapproved 1979). In: Annual book of ASTM standards, Part 40. Philadelphia, Pennsylvania, pp. 870-876.

American Society for Testing and Materials. 1981a. Standard method for polychlorinated biphenyls (PCBs) in water. ANSI/ASTM D 3534-80. In: Annual book of ASTM standards, Part 31. Philadelphia, Pennsylvania, pp. 816-833.

American Society for Testing and Materials. 1981b. Standard method for analysis of environmental materials for polychlorinated biphenyls. ANSI/ASTM D 3304-77. In: Annual book of ASTM standards, Part 31. Philadelphia, Pennsylvania, pp. 877-885.

American Society for Testing and Materials. 1983. Standard method for analysis of polychlorinated biphenyls in mineral insulating oils by gas chromatography. ANSI/ASTM D 4059-S3. In: Annual book of ASTM standards, Part 40. Philadelphia, Pennsylvania, pp. 542-550.

Association of Official Analytical Chemists. 1980a. General method for organochloride and organophosphorus pesticides, Method 29.001. Official

Methods of Analysis of the Association of Official Analytical Chemists, W. Horwitz, Ed. (13th ed., Washington, DC), pp. 466-474.

Association of Official Analytical Chemists. 1980b. PCB in paper and paperboard, Method 29.029. Official Methods of Analysis of the Association of Official Analytical Chemists, W. Horwitz, Ed. (13th ed., Washington, DC), p. 475-476.

Ballinger DG. 1978 (December 11). Test procedures for priority organics in municipal wastewater and sludges. U.S. Environmental Protection Agency, Cincinnati, Ohio.

Beard JH III, Schaum J. 1978 (February 10). Sampling methods and analytical procedures manual for PCB disposal: Interim Report, Revision 0. Office of Solid Waste, U.S. Environmental Protection Agency, Washington, DC.

Bellar TA, Lichtenberg JJ. 1982. The determination of polychlorinated biphenyls in transformer fluid and waste oils. Prepared for U.S. Environmental Protection Agency, EPA-600/ 4-81-045.

Boe B, Egaas E. 1979. Qualitative and quantitative analyses of polychlorinated biphenyls by gas-liquid chromatography. J Chromatogr 180:127-132.

Brinkman UATh, De Kok A, De Vries G, Reymer HGM. 1976. High-speed liquid and thin-layer chromatography of polychlorinated biphenyls. J Chromatogr 128:101-110.

Bush B, Baker F, Dell'Acqua R, Houck CL, Lo F-C. 1975. Analytical response of polychlorinated biphenyl homologues and isomers in thin-layer and gas chromatography. J Chromatogr 109:287-295.

Bush B, Connor S, Snow J. 1982. Glass capillary gas chromatography for sensitive, accurate polychlorinated biphenyl analysis. J Assoc Off Anal Chem 65(3):555-566.

Colby BN, Burns EA, Lagus PL. 1983. The S-Cubed PCBA 101, an automated field analyier for PCBs. Abstract No. 731, 1983 Pittsburgh Conference and Exposition on Analytical Chemistry and Applied Spectroscopy.

Collins GB, Holmes DC, Jackson FJ. 1972. The estimation of polychlorobiphenyls. J Chromatogr 71:443-449.

Copland GB, Gohmann CS. 1982. Improved method for polychlorinated biphenyl determination in complex matrices. Environ Sci Technol 16:121-124.

De Vos RH. 1972. Analytical techniques in relation to the contamination of the fauna. TNO-nieuws 27:615-622.

De Vos RH, Peet EW. 1971. Thin-layer chromatography of polychlorinated biphenyls. Bull Environ Contam Toxicol 6(2):164-170.

Devenish I, Harling-Bowen L. 1980. The examination and estimation of the performance characteristics of a standard method for organo-chlorine insecticides and PCB. In: Hydrocarbons and Halogenated Hydrocarbons in the Aquatic Environment, B. K. Afghan and D. Mackay, Eds. (New York: Plenum Press), pp. 231-253.

Dow Chemical Company. 1981 (July 1). Determination of chlorinated biphenyls in the presence of chlorinated benzenes. Midland MI.

Dry Color Manufacturers Association. 1981. An analytical procedure for the determination of polychlorinated biphenyls in dry phthalocyanine blue, phthalocyanine green, and diarylide yellow pigments. Arlington, VA.

Erickson MD, Stanley JS, Turman K, Radolovich G, Bauer K, Onstot J, Rose D, Wickham M. 1982. Analytical methods for by-product PCBs--preliminary validation and interim methods. Interim Report No 4, Office of Toxic Substances, U.S. Environmental Protection Agency, Washington, D. C., EPA-560/5-82-006, NTIS No. PB83 127 696, 243 pp.

Erickson MD, Stanley JS, Radolovich G, Blair RB. 1983 (August 15). Analytical method: the analysis of by-product chlorinated biphenyls in commercial products and product wastes. Revision 1, Prepared by Midwest Research Institute for Office of Toxic Substances, U.S. Environmental Protection Agency, Washington, DC, under Subcontract No. A-3044(8149)-271, Work Assignment No. 17 to Battelle, Washington, DC.

Erickson MD 1984a. Analytical method: The analysis of by-product chlorinated biphenyls in commercial products and product wastes, revision 2.

U.S. Environmental Protection Agency, Office of Toxic Substances, Washington, DC, EPA 560/5-85-010.

Erickson MD. 1984b. Analytical method: The analysis of by-product chlorinated biphenyls in water, revision 2. U.S. Environmental Protection Agency, Office of Toxic Substances, Washington, DC, EPA 560/5-85-012.

Erickson MD. 1985. The analytical chemistry of PCBs. Butterworths, Boston, MA.

Erickson MD, Stanley JS, Turman JK, and Radolovich G. 1985a. Analytical method: The analysis of chlorinated biphenyls in liquids and solids. U.S. Environmental Protection Agency, Office of Toxic Substances, Washington, DC, EPA-560/5-85-023.

Fisher DJ, Rouse TO, Lynn TR. 1984. Field determination of PCB in transformer oil "CLOR-N-OIL Kit." In: Proceedings: 1983 PCB Seminar, Addis G, and Komai RY, Eds. Report No. EPRI-EL-3581, Palo Alto, CA: Electric Power Research Institute.

Food and Drug Administration. Pesticide Analytical Manual. Vol. I, August 1, 1977.

Haile CL, Baladi E. 1977. Methods for determining the total polychlorinated biphenyl emissions from incineration and capacitor and transformer filling plants. U.S. Environmental Protection Agency, EPA-600/4-77-048, NTIS No. PB-276 745/7G1.

Haile CL, Lopez-Avila V. 1984. Development of analytical test procedures for the measurement of organic priority pollutants--project summary. U.S. Environmental Protection Agency, Environmental Monitoring and Support Laboratory, Cincinnati, Ohio, EPA-600/54-84-001; (Full Report available as NTIS No. PB 84-129 048).

Harris RW, Grainger CF, Mitchell WJ. 1981. Validation of a method for measuring polychlorinated biphenyls in natural gas pipelines. EPA 600/4-81-048; NTIS No. PB82-207556.

Hattori Y, Kuge Y, Nakamoto M. 1981. The correlation between the electroncapture detector response and the chemical structure for polychlorinated biphenyls. Bull Chem Soc Jpn 54(9):2807-2810; Chem Abstr 96:34427s (1981).

Kan T, Kamata K, Ueta T, Yamazoe R, Totani T. 1973. Fluorescence reactions of organohalogen compounds. I. Fluorometry of polychlorinated biphenyls (PCB) with diphenylamine on thin-layer chromatograms. Tokyo Toritsu Eisei Kenkyusho Kenky Nempo 24:137-145; Chem Abst 80:115771w (1974).

Kawabata J. 1974. Simple method for the determination of PCBs [polychlorinated biphenyls] by a combination of thin-layer chromatography and UV absorption. Kogai To Taisaku 10(10):1112-1116; Chem Abst 83:201652b (1975).

Koch R. 1979. Circular thin-layer chromatography as a rapid method for a qualitative detection of organochlorine compounds. Acta Hydrochim Hydrobiol 7(3):355-356; Chem Abst 91:1015742 (1979).

Levine SP, Homsher MT, Sullivan JA. 1983. Comparison of methods of analysis of polychlorinated biphenyls in oils. J Chromatogr 257:255-266.

Levins PL, Rechsteiner CE, Stauffer JL. 1979. Measurement of PCB emissions from combustion sources. U.S. Environmental Protection Agency, Report No. EPA-600/7-79-047.

Lewis RG. 1982 (March). Procedures for sampling and analysis of polychlorinated biphenyls in the vicinities of hazardous waste disposal sites. U.S. Environmental Protection Agency, Research Triangle Park, NC, 14 pp.

Lingle JW. Wisconsin Electric Power Company, P.O. Box 2046, Milwaukee, WI 53201. May 24, 1985. Personal communication.

Longbottom JE, Lichtenberg JJ, Eds. 1982 (July). Methods for organic chemical analysis of municipal and industrial wastewater. U.S. Environmental Protection Agency, Report No. EPA-600/4-82-057.

Lovett AM, Nacson S, Hijazi NH, Chan R. 1983. Real time ambient air measurements for toxic chemical. In: Proceedings: a specialty conference on: measurement and monitoring of non-criteria (toxic) contaminants in air, Frederick ER, Ed., The Air Pollution Control Association, Pittsburgh, PA, 113-125 pp.

Mason BJ. 1982 (October). Preparation of soil sampling protocol: techniques and strategies. ETHURA, McLean, VA, under subcontract to Environmental Research Center, University of Nevada, for U.S. Environmental Protection Agency, Las Vegas.

Matern B. 1960. Spacial variation. Medd. fr. Stateus Skogsforsknings Institut. 49:1-144.

McQuade JM. 1982. PCB analysis by X-ray fluorescence. In: Proceedings: 1981 PCB Seminar, Addis G, Marks J, Eds., Report No. EPRI-EL-2572, Palo Alto, CA: Electric Power Research Institute, pp. 2-9.

Millar JD, Thomas RE, Schattenberg HJ. 1984 (June). EPA Method Study 18, Method 608--organochlorine pesticides and PCB's. Quality Assurance Branch, Environmental Monitoring and Support Laboratory, U.S. Environmental Protection Agency, Cincinnati, Ohio. Report No. EPA-600/4-84-061, NTIS No. PB84 211358, 197 pages.

Moein GJ. 1976. Study of the distribution and fate of polychlorinated biphenyls and benzenes after spill of transformer fluid. Report No. EPA 904/9-76-014, NTIS No. PB288484.

Mulhern BM. 1968. An improved method for the separation and removal of organochlorine insecticides from thin-layer plates. J Chromatogr 34:556-558.

Mulhern BM, Cromartie E, Reichel WL, Belisle A. 1971. Semiquantitative determination of polychlorinated biphenyls in tissue samples by thin layer chromatography. J Assoc Offic Anal Chem 54(3):548-550.

National Institute for Occupational Safety and Health. 1977a (April). NIOSH Manual of Analytical Methods, Second Edition, Part I, NIOSH Monitoring Methods, Vol. 1, "Polychlorinated Biphenyls (PCB) in Air, Analytical Method P&CAM 244," U.S. Department of Health, Education, and Welfare, Cincinnati, Ohio.

National Institute for Occupational Safety and Health. 1977b (April). NIOSH Manual of Analytical Methods, Second Edition, Part I, NIOSH Monitoring Methods, Vol. 1, "Polychlorinated Biphenyls (PCB) in Air, Analytical Method P&CAM 253," U.S. Department of Health, Education, and Welfare, Cincinnati, Ohio.

NIOSH. 1977c (September). National Institute for Occupational Safety and Health. Criteria for a recommended standard.. ...occupational exposure to poly-chlorinated biphenyls (PCBs). U.S. Department of Health, Education, and Welfare (Public Health SerYice, Center for Disease Control, and National Institute for Occupational Safety and Health), DHEW (NIOSH) Publication No. 7-225, 224 pp.

NIOSH. 1980 (September). National Institute for Occupational Safety and Health, U.S. Department of Health and Human Services. Health Hazard Evaluation Report No. 80-85-745. Oakland, CA: Pacific Gas and Electric Company.

Onsuka FI, Kominar RJ, Terry KA. 1983. Identification and determination of polychlorinated biphenyls by high-resolution gas chromatography. J Chromatogr 279:111-118.

Picker JE, Colby BN. 1984. Field determination of Aroclors using an automated electron capture detector gas chromatograph. In: Proceedings: 1983

PCB Seminar, Addis G, Komai RY, Eds., Report No. EPRI-EL-3581. Palo Alto, CA: Electric Power Research Institute.

Piechalak B. 1984. The semiquantitative detection of polychlorinated biphenyls (PCBs) in contaminated soils by thin-layer chromatography. In: Proceedings: 1983 PCB Seminar, Addis G, Komai RY, Eds., Report No. EPRI-EL3581. Palo Alto, CA: Electric Power Research Institute.

Rodriguez CF, McMahon WA, Thomas RE. 1980 (March). Method development for determination of polychlorinated hydrocarbons in municipal sludge. Final Report, Contract No. 68-03-2606, Environmental Protection Agency, EPA-600/2-80-029; NTIS No. PB 82-234 071.

Sawyer LD. 1973. Collaborative study of the recovery and gas chromatographic quantitation of polychlorinated biphenyls in chicken fat and polychlorinated biphenyl-DDT combinations in fish. J Assoc Offic Anal Chem 56(4):1015-1023.

Sawyer LD. 1978. Quantitation of polychlorinated biphenyl residues by electron capture gas-liquid chromatography: reference material characterization and preliminary study. J Assoc Offic Anal Chem 61(2):272-281.

Schwalb AL, Marquez A. 1982. Salt River Project's experience with the Horiba Sulfur/Chlorine-in-Oil Analyzer. In: Proceedings: 1981 PCB Seminar, Addis G, Marks J, Eds., Report No. EPRI-EL-2572. Palo Alto, CA: Electric Power Research Institute, pp. 2-23.

Sherma, J. Manual of Analytical Quality Control for Pesticides and Related Compounds in Human and Environmental Samples, EPA-600/2-81-059; NTIS No. PB81-222721 (April 1981).

Singer E, Jarv T, Sage M. 1983. Survey of polychlorinated biphenyls in ambient air across the province of Ontario. Chapter 19 in Physical Behavior

of PCBs in the Great Lakes, Mackay D, Paterson S, Eisenreich SJ, Simmons MS, Eds. Ann Arbor, MI: Ann Arbor Science Publishers, Inc., pp 367-383.

Sonchik S, Madeleine D, Macek P, Longbottom J. 1984. Evaluation of sample preparation techniques for the analysis of PCBs in oil. J Chromatogr Sci 22:265-271.

Spittler TM. 1983. Field measurement of PCB's in soil and sediment using a portable gas chromatograph. Natl Conf Manage Uncontrolled Hazard Waste Sites 105-107; Chem Abst 100:220890p (1984).

Stalling DL, Huckins JN. 1973. Reverse phase thin layer chromatography of some Aroclors, halowaxes, and pesticides. J Assoc Offic Anal Chem 56(2): 367-372.

Stahr HM. 1984. Analysis of PCBs by thin layer chromatography. J Liq Chrom 7(7):1393-1402.

Tahiliani VH. 1984. CLOR-N-OIL field test program. In: Proceedings: 1983 PCB Seminar, Addis G, Komai RY, Eds., Report No. EPRI-EL-3581. Palo Alto, CA: Electric Power Research Institute.

Tanabe H. 1976. PCB microanalysis. In PCB Poisoning and Pollution, K. Higuchi, Ed. (Tokyo: Kodansha, Ltd; New York: Academic Press), pp. 127-145.

Tuinstra LGMTh, Driessen JJM, Keukens HJ, Van Munsteren TJ, Roos AH, Traag WA. 1983. Quantitative determination of specified chlorobiphenyls in fish with capillary gas chromatography and its use for monitoring and tolerance purposes. Intern J Environ Anal Chem 14:147-157.

Ueta T, Kamata K, Kan T, Kazama M, Totani T. 1974. Fluorescence reactions for organic halogen compounds. II. In situ fluorometry of polychlorinated biphenyls and their isomers on thin-layer chromatograms using diphenylamine. Tokyo Toritsu Eisei Kenkyusho Kenkyu Nempo 25:111-118; Chem Abst 83:21991c (1975).

United Kingdom Department of the Environment. 1979. Organochlorine Insecticides and Polychlorinated Biphenyls in Waters 1978; Tentative Method. Methods for the Examination of Waters and Associated Materials. Organochlorine Insectic. Polychlorinated Biphenyls Waters 28 pp.

USEPA. 1978 (September). U.S. Environmental Protection Agency. Methods for benzidine, chlorinated organic compounds, pentachlorophenol and pesticides in water and wastewater. Interim Report, Environmental Monitoring and Support Laboratory, Cincinnati, OH.

USEPA. 1979a (December 3). U.S. Environmental Protection Agency. Organo-chlorine pesticides and PCBs--Method 608. 44 FR 69501-69509.

USEPA. 1979b (December 3). U.S. Environmental Protection Agency. Base/neutrals, acids, and pesticides--Method 625. 44 FR 69540-69552.

USEPA. 1979c (September). U.S. Environmental Protection Agency. Analytical protocol for screening publicly owned treatment works (POTW) sludges for organic priority pollutants. Environmental Monitoring and Support Laboratory, Cincinnati, OH.

USEPA. 1980. U.S., Environmental Protection Agency. Guidelines and specifications for preparing quality assurance project plans. Office of Monitoring Systems and Quality Assurance, QAMS-005/80.

USEPA. 1981a (February). U.S. Environmental Protection Agency. The analysis of polychlorinated biphenyls in transformer fluid and waste oils. Office of Research and Development, Environmental Monitoring and Support Laboratory, Cincinnati, OH.

USEPA. 1981b. U.S. Environmental Protection Agency. PCB disposal by thermal destruction. Solid Waste Branch, Air and Hazardous Materials Division, Region 6, Dallas, TX, EPA-200/9-81-001; NTIS No. PB82 241 860, 606 pp.

USEPA. 1981c (March). U.S. Environmental Protection Agency. TSCA Inspection Manual.

USEPA. 1982a (November 4). U.S. Environmental Protection Agency. Analysis of pesticides, phthalates, and polychlorinated biphenyls in soils and bottom sediments. HWI Sample Management Office, Alexandria, VA, unpublished method, 12 pp.

USEPA. 1982b (July). U.S. Environmental Protection Agency. Test methods for evaluating solid waste, physical/chemical methods, SW-846, 2nd ed. Office of Solid Waste and Emergency Response, Washington, DC.

USEPA. 1982c (October 21). 40 CFR Part 761, Polychlorinated Biphenyls (PCBs); Manufacturing, Processing, Distribution in Commerce, and Use Prohibitions; Use in Closed and Controlled Waste Manufacturing Processes. Fed. Reg. 47:46980-46986.

USEPA. 1982d (November 4). Analysis of Pesticides, Phthalates, and Polychlorinated Biphenyls in Soils and Bottom Sediments. HWI Sample Management Office, Alexandria, VA, unpublished method, 12 pp.

USEPA. 1982e (July). Test Methods for Evaluating Solid Waste-Physical/Chemical Methods, SW-846, 2nd Edition. Office of Solid Waste and Emergency Response, Washington, DC.

USEPA. 1983. U.S. Environmental Protection Agency. Quality assurance program plan for the Office of Toxic Substances, Office of Pesticides and Toxic Substances, Washington, D.C.

USEPA. 1984a (October 20). Organochlorine Pesticides and PCBs--Method 608. Fed. Reg. 49(209):89-104.

USEPA. 1984b (October 26). Base/Neutrals, Acids, and Pesticides--Method 625. Fed. Reg. 49(209):153-174.

USEPA. 1984-c (October 11). 4-0 CFR Part 761, Polychlorinated Biphenyls (PCBs); Manufacture, Processing, Distribution in Commerce and Use Prohibitions; Use in Electrical Transformers. Fed. Reg. 49:39966-39989.

USEPA. 1984d (October). Mass Spectrometric Identification and Measurement of Polychlorinated Biphenyls as Isomer Groups. Draft Report by Physical and Chemical Methods Branch, Office of Research and Development, Cincinnati, OH.

USEPA. 1984e (July 10). 40 CFR Part 761, Polychlorinated Biphenyls (PCBs); Manufacturing, Processing, Distribution in Commerce and Use Prohibitions; Response to Individual and Class Petitions for Exemptions, Exclusions, and Use Authorization, Final Rule. Fed. Reg. 49:28154-28209.

USWAG. 1984 (October 15). The Utility Solid Waste Activities Group. Proposed spill cleanup policy and supporting studies. U.S. Environmental Protection Agency.

Watts RR (Ed.). 1980 (June). Analysis of Pesticide Residues in Human and Environmental Samples, A Compilation of Methods Selected for Use in Pesticide Monitoring Programs, U.S. Environmental Protection Agency, Research Triangle Park, NC, EPA-600/8-80-038.

Webb RG, McCall AC. 1973. Quantitative PCB standards for electron capture gas chromatography. J Chromatogr Sci 11:366-373.

Willis DE, Addison RF. 1972. Identification and estimation of the major components of a commercial polychlorinated biphenyl mixture, Aroclor 1221. J Fish Res Board Can 29(5):592-595.

Zell M, Ballschmiter K. 1980. Baseline study of the global pollution. III. Trace analysis of polychlorinated biphenyls (PCB) by ECD glass capillary gas chromatography in environmental samples of different trophic levels. Fresenius' Z Anal Chem 304:337-349.

Zitko V, Hutzinger O, Safe S. 1971. Retention times and electron-capture detector responses of some individual chlorobiphenyls. Bull Environ Contam Toxicol 6(2):160-163.

Part II

Field Manual for Grid Sampling of PCB Spill Sites to Verify Cleanup

Gary L. Kelso and **Mitchell D. Erickson**, Midwest Research Institute
David C. Cox, Washington Consulting Group

I. Scope and Application

The purpose of this manual is to provide detailed, step-by-step guidance to EPA staff for using hexagonal grid sampling at a PCB spill site. Emphasis is placed on sampling sites which have already been cleaned, although the sampling methods presented may also be used at PCB spill sites which have not been cleaned. Guidance is given for preparing the sample design; collecting, handling, and preserving the samples taken; maintaining quality assurance and quality control; and documenting and reporting the sampling procedures used. An optional strategy for compositing samples is given in the appendix.

This is a companion document to the report "Verification of PCB Spill Cleanup by Sampling and Analysis" (EPA 560/5-85-026, August 1985, Second Printing). That report provides an overview of PCB spill cleanup activities and guidelines for sampling and analysis including: sampling designs, sampling techniques, analytical techniques, selection of appropriate analytical methods, quality assurance, documentation and records, and reporting results. The previous report provided the rationale and background for the techniques selected and describes many options in greater detail.

This "how-to" report concentrates on detailed guidance for field sampling personnel and does not attempt to provide background information on the techniques presented. This manual addresses field sampling only and does not provide information on laboratory procedures, including sample analysis, data reduction and laboratory data reporting. The types of field sampling situations discussed in this manual are those typically found when a PCB spill results from a PCB article, PCB container, or PCB equipment spill. Unusual PCB

0-87371-945-X/93/0.00 + $.50

spill situations, such as elongated spills on highways from a moving vehicle, large spills in waterways, and large, catastrophic spills, are not addressed.

II. Summary

This manual is divided into the following sections:

- •Safety
- •Sampling Equipment and Materials
- •Sample Design
- •Sample Collection, Handling, and Preservation Quality Assurance
- •Quality Control
- •Documentation and Records
- •Validation of the Manual

Safety aspects of field sampling include wearing proper protective equipment, practicing good hygiene, using safe work practices, and training field inspectors in safety procedures. Sampling equipment and materials include personnel equipment, sampling equipment, and documentation materials. Prior to making the field sampling trip, the EPA inspector should ensure that all sampling equipment and materials are available, and that all sampling containers and equipment have been properly precleaned.

The sample design is based on a hexagonal grid of 7, 19, or 37 sample points. A step-wise method describes how to construct a diagram of the PCB spill site on graph paper; determine the radius and center of the sampling circle; determine which grid size to use; lay out the grid on the diagram; and then lay out the sampling grid on the site.

After the sampling grid has been laid out on the site, a sample must be taken at each grid point. Methods to collect, handle, and preserve different types of samples, including surface soil samples, soil core samples, surface and subsurface water samples, wipe samples from nonporous hard surfaces, destructive samples from porous hard surfaces, and vegetation samples, are suggested. For each type of sample to be taken, methods are recommended to prevent cross-contamination between samples.

Quality assurance (QA) and quality control (QC) must be an integral part of any sampling scheme. A quality assurance plan must be developed by appropriate EPA offices according to EPA guidelines and be submitted to the regional QA officer or other appropriate QA official for approval prior to sampling PCB spill sites. Each EPA office must operate a formal QC program and all QC measures should be stipulated in the QA plan. Some of the requirements of quality control are discussed in this report, including field blanks, sampling without cross-contamination, sample custody, and documentation of the field sampling activities.

All sampling activities should be thoroughly documented and reported as a part of the verification process. Each EPA office is responsible for preparing and maintaining complete records, including an equipment preparation log book, a field log book, site description forms, chain-of-custody forms, sample analysis request forms, and field trip reports.

Section X briefly describes a field study which was conducted to test and validate the sample design given in this manual. The study showed that the sampling design is easy to follow and understood by those unfamiliar with the manual prior to reading it, and that the grid sample points can be correctly laid out in a relatively short period of time.

The appendix gives strategies that may be used to composite the samples taken at a PCB spill site when compositing is deemed to be desirable.

III. Safety

A PCB spill site which has been cleaned up should have very low levels of PCBs present. The EPA inspector(s) who sample the site to verify that the site has been properly cleaned up should, however, take some precautions to minimize any exposure to PCBs or other potential hazards at the site.

In order to ensure that the inspectors understand and practice good safety procedures, a training and education program should be established and a health and safety manual provided by the responsible EPA officer. The program should inform inspectors of the potential hazards of exposure to PCBs, and the proper safety procedures to follow when sampling PCB spill sites.

IV. Sampling Equipment and Materials

The equipment and materials required to sample a PCB spill site will vary with the types of samples to be taken. The general lists of equipment and materials given below must be adjusted for the specific requirements of each spill. The lists include personnel equipment, sampling equipment and materials, and documentation materials which should be taken to the spill site by the EPA inspector. These equipment and materials must be assembled prior to making the site visit, and all sampling containers and sampling equipment must be precleaned.

A. Personnel Equipment

The inspector should take the following personnel equipment to the spill site:

- Disposable rubber gloves
- Plastic overshoes
- Safety glasses
- Impervious paper-like coveralls
- Hardhat
- Safety shoes
- First-aid kit
- Other safety equipment specified by safety officer

B. Sampling Equipment and Materials

Since the types of samples to be taken at a spill site may vary from site to site, the following sampling equipment and materials should be taken:

- Precleaned glass sample jars with Teflon-lined caps
- Aluminum foil (solvent-rinsed)
- Container of reagent-grade solvent (isooctane is recommended)
- Box of 11 cm filter paper (e.g., Whatman 40 ashless or Whatman 50 smear tabs)
- Gauze pads
- Stainless steel forceps
- Stainless steel templates (10 cm x 10 cm square)
- Stainless steel trowels, Teflon scoops, or laboratory spatulas (precleaned)
- Soil coring devices (such as King-tube samplers or piston corers) Hammer and chisel
- Hole saw and drill
- Pruning shears
- Stainless steel buckets
- Disposable wiping cloths
- Plastic disposable bags
- Sample bags and seals
- Survey stakes
- 100 ft tape measure
- Ice chests containing ice or ice packs and secured with padlocks Compass and maps
- Duct tape
- Subsurface water sampling equipment (such as pumps, siphons, glass

sampling jars with attachments, etc.)
- •Container of distilled water
- •Stainless steel mixing bowls and spoons

C. Documentation Materials

The following documentation materials should be taken to the field site:

- •Field log book
- •Chain-of-custody forms
- •Site description forms
- •Sample analysis request forms
- •Sample bottle labels
- •Camera with film
- •Yellow TSCA PCB marks

D. Trip Preparation

The EPA field inspector must assemble all the necessary equipment and materials prior to making the field sampling trip. Special attention should be given to assuring that all of the equipment and materials are available, and that the sample containers and sampling equipment have been properly precleaned. The equipment preparation should be documented in a log book (Section IX.A) prior to making the trip.

V. Sample Design

The methods to be used for determining the sample point locations at a PCB spill site are given in this section, and are based upon a hexagonal grid sample design which was recommended in the report "Verification of PCB Spill Cleanup by Sampling and Analysis." Although the grid design involves more samples and a more complicated layout than the usual grab sampling methods, the grid design is essential to obtaining a representative sample of the site and greatly increases the chance of detecting high levels of PCB contamination when they exist. For example, when 4% of the PCB spill site remains contaminated at 50 ppm after cleanup, analysis of samples from a 37-point grid has a 98% chance of detection of this contamination level, while analysis of six random grab samples from the site has only a 3% chance of detection (Boomer et al. 1985).

The hexagonal grid sampling design is to be laid out within a sample circle centered on the spill site, and extending just beyond its boundaries. Preparation of the design requires the following steps:

Step 1: Diagram the Cleanup Site
Step 2: Diagram All Cleanup Surfaces in the Same Plane
Step 3: Find the Center and Radius of the Sampling Circle
Step 4: Determine the Number of Grid Sample Points to Use
Step 5: Lay Out the Sampling Points on the Diagram Constructed in Step 2
Step 6: Lay Out the Sampling Locations on the Site
Step 7: Consider Special Cases and Use Judgment for Sample Points

The discussion which follows gives the methods to be used in accomplishing each step of the hexagonal grid sampling design, using three dimensional spill surface as an example. Following this discussion, a simple example of laying out the sample design on a rectangular two-dimensional surface is given.

A. Step 1: Diagram the Cleanup Site

Draw a scale diagram of the cleanup site on graph paper, including vertical surfaces (walls, fences, etc.), noting important dimensions and different types of surfaces (sod, cement, asphalt, etc.). Such a diagram may sometimes be found in records of the cleanup. If not, site measurements should be taken. Great accuracy (e.g., using surveying instruments) is not necessary, however; the use of a tape measure and pacing should be adequate. An example diagram is shown in Figure 1 on a scale of 1 in. = 4 ft. The site diagram should include as many reference points as necessary to relocate the spill area in the future, if necessary. For example, a spill site in an open field should be located with respect to nearby structures such as roads, telephone poles, buildings, etc. The direction of north should be indicated on the diagram.

If available, a detailed drawing or a survey plot of the spill site should be obtained from the individual(s) that cleaned the site.

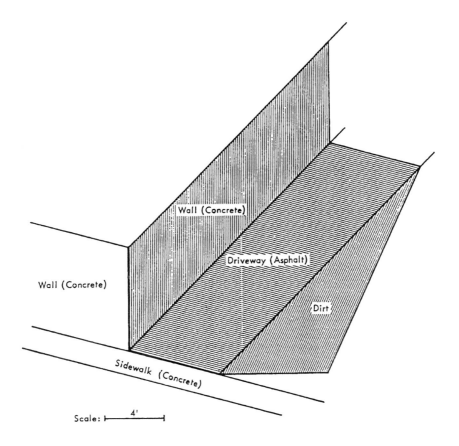

Figure 1. Example PCB spill site diagram.

B. Step 2: Diagram All Cleanup Surfaces in the Same Plane

The purpose of this second diagram is to determine and show the dimensions of the total cleanup area, including vertical surfaces, so that the required sample size can be found.   The diagram also facilitates the determination of sampling locations on vertical surfaces.   Constructing the diagram is analogous to flattening a cardboard box.   All vertical surfaces are placed in the same plane as the adjoining horizontal surfaces.   Figure 2, also on a scale of 1 in. = 4 ft, shows the example spill cleanup site diagrammed in the same plane. The actual site dimensions are shown in feet.

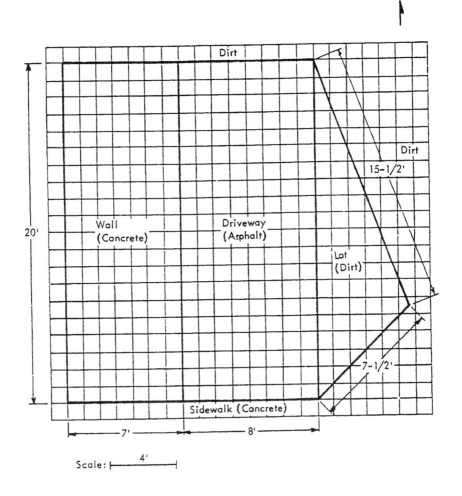

Figure 2. Example spill cleanup site diagrammed in the same plane.

C. Step 3: Find the Center and Radius of the Sampling Circle

In practice, the contaminated area from a spill will be irregular in shape. In order to standardize sample design and layout in the field, samples are collected within a circular area surrounding the contaminated area. The sampling circle is, approximately, the smallest circle containing all cleanup surfaces diagrammed in Step 2.

A recommended procedure for finding the center and radius of the sampling circle is illustrated in Figure 3 and is described below:

1. Draw the longest dimension, L_1, of the site diagram in Step 2.
2. Find the midpoint, P, of L_1.
3. Draw a second dimension, L_2, through P perpendicular to L_1. L_2 extends to the boundaries of the site diagram.
4. The midpoint, C, of L_2 is the <u>center</u> of the sampling circle.
5. The distance from C to either end of the longest dimension, L_1, is the <u>sampling radius</u>, r.

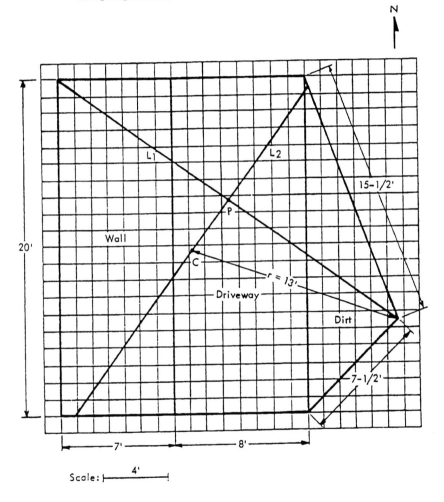

Figure 3. Locating the center and sampling radius of the example spill cleanup site.

Figure 4 illustrates the application of this procedure to a site with an irregular shape, and Figure 5 shows the procedure for a variety of irregularly shaped areas. These figures show that the center and radius determined are generally reasonable.

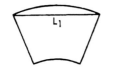

(a) Draw longest dimension, L_1, on site diagram.

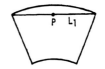

(b) Find midpoint, P, of L_1.

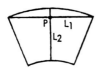

(c) Draw line, L_2, through P perpendicular to L_1.

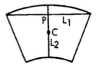

(d) The midpoint, C, of L_2 is the center of the sampling circle.

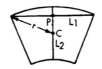

(e) The distance from C to the end of L_1 is the sampling radius, r.

Figure 4. Method to find center and radius of the sampling circle.

D. Step 4: Determine the Number of Grid Sample Points to Use

The number of grid samples to be taken at a site depends upon the radius of the sampling circle, which is determined from the scale diagram shown in Figure 3. The number of samples to be taken at a spill site should increase as the radius of the sample circle increases. The reason for this is that the probability of detecting residual PCB contamination at a given site increases as the number of grid samples increases. Table 1 shows the required number of grid samples for sampling circles with a radius of 4 ft or less (seven samples); greater than 4 ft to 11 ft (19 samples); and greater than 11 ft (37 samples).

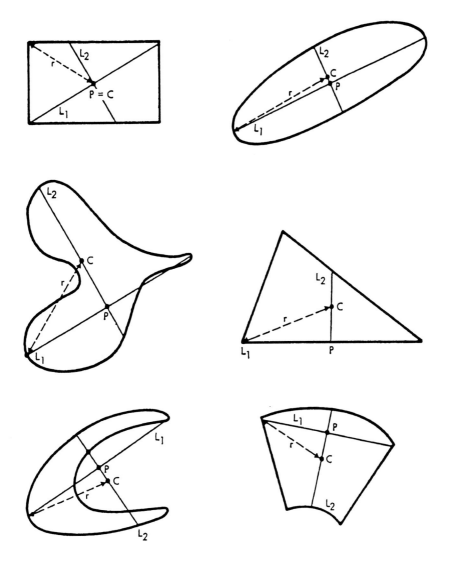

Figure 5. Locating the center and sampling circle radius of irregularly shaped spill areas.

Table 1. Required Number of Grid Samples Based on the Radius of the Sampling Circle

Sampling radius, r (ft)	Number of Samples
≤ 4	7
> 4 - 11	19
> 11	37

The radius, r, for the example site is 3-1/4 in. in Figure 3. Thus, the actual site sampling radius is 13 ft (3-1/4 in. x 4 ft/in.) and the number of grid samples required is 37.

Figures 6, 7, and 8 illustrate the hexagonal grid sampling design for the three sample sizes given in Table 1, for a sampling radius of 4, 10, and 20 ft, respectively.

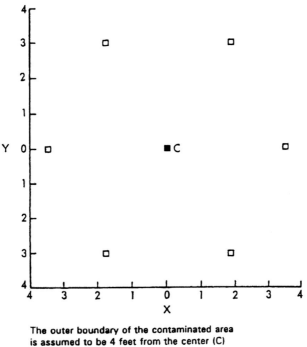

The outer boundary of the contaminated area is assumed to be 4 feet from the center (C) of the spill site.

Figure 6. Location of sampling points in a 7-point grid.

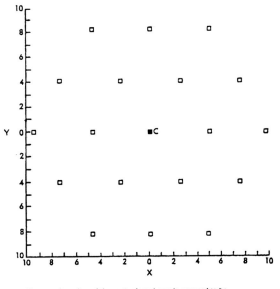

The outer boundary of the contaminated area is assumed to be
10 feet from the center (C) of the spill site.

Figure 7. Location of sampling points in a 19-point grid.

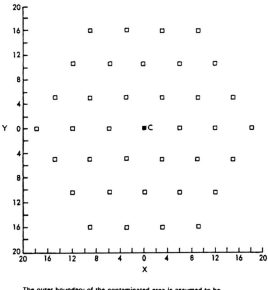

The outer boundary of the contaminated area is assumed to be
20 feet from the center (C) of the spill site.

Figure 8. Location of sampling points in a 37-point grid.

E. Step 5: Lay Out the Sampling Points on the Diagram Constructed in Step 2

The geometric properties of the hexagonal designs can be used in many ways to lay out the sampling points. Perhaps the simplest way to proceed is as follows. Define s to be the distance between adjacent points and u to be the distance between successive rows of the design. The distances s and u are given in terms of the sampling radius, r, in Table 2 below for the given number of samples defined by the radius rule and listed in Table 1.

Table 2. Geometric Parameters of the Hexagonal Grid Designs for Sampling Radius r

Number of samples	Distance, s, between adjacent sample points	Distance, u, between successive rows
7	0.87r	0.75r
19	0.48r	0.42r
37	0.30r	0.26r

The recommended method for laying out the sample points of the hexagonal grid on the scale diagram is illustrated in Figure 9 and is described below.

1. Draw a diameter of the sampling circle on the scale diagram. The orientation of the diameter (e.g., east-west) should be chosen to maximize the number of sample points which fall within the spill area, when practical. A transparent overlay like Figures 6, 7 and 8 (using the appropriate scale) may be helpful in determining the orientation of the diameter.

2. Place the center point of the hexagonal design at the center (C) of the sampling circle. Lay out the middle row of the design along the diameter with successive points a distance, s, apart.

3. To lay out the next row, find the midpoint between the last two sample points of the middle row and move a distance, u, perpendicular to the middle row as shown in Figure 9. This is the first sample point of the next row. Now lay out the remaining points at distance s from each other. By systematically following this plan, the entire design can be laid out.

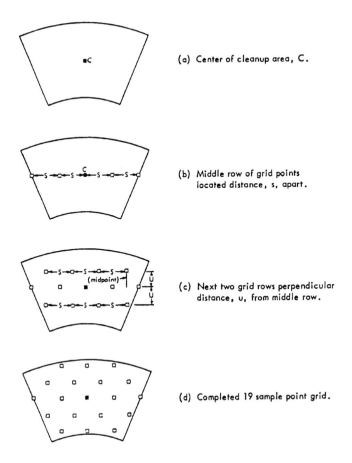

(a) Center of cleanup area, C.

(b) Middle row of grid points located distance, s, apart.

(c) Next two grid rows perpendicular distance, u, from middle row.

(d) Completed 19 sample point grid.

Figure 9. Construction of sampling grid on a site diagram.

Figure 10 shows the sample point locations for the 37 grid points for the example PCB spill site diagrammed previously in Figures 1, 2, and 3. On the diagram, $r = 3\text{-}1/4$ in. so from Table 2 the grid spacing is $s = 0.30r = 1$ in. and the distance between the rows is $u = 0.26r = 7/8$ in.

In Figure 10, a horizontal diameter is drawn through C. Sampling locations 1 through 7 are marked 1 in. apart. To lay out the next row of the design, we first find location 8. Point D is the midpoint between locations 3 and 4. Then, as ascribed, location 8 is a vertical distance $u = 7/8$ in. (3 ft 6 in. on the site) above D. Now locations 9 through 13 are laid out 1 in. apart. In the same way, locations 14 through 18 are found. Continuing so, the entire grid is marked on the diagram.

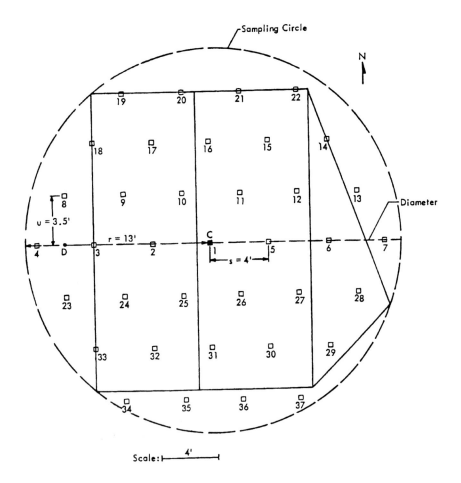

Figure 10. Sampling locations on the example PCB spill site.

All of the sample points in Figure 10 are numbered (1 to 37). Any type of numbering system can be used, but the points must each be identified so that the location of the samples taken can be identified by reference to the diagram points.

Note that sampling locations 4, 7, 8, 13, 23, 34, 35, 36, and 37 are outside the cleanup area. Of these, locations 4, 8, 23, 34, and 35 do not correspond to a physical location--all are in "thin air," so to speak--and samples cannot be collected at these locations. Locations 36 and 37 are concrete samples; locations 7 and 13 are dirt samples (from Figure 2).

The orientation of the sample circle diameter shown does not actually maximize the number of points falling within the spill area, since a 45° clockwise rotation would result in only 8 points lying outside the spill area instead of the 9 points shown. However, a 45° orientation would make the sample points very difficult to locate on the actual site with little to gain by the addition of one more sample point within the spill area.

F. Step 6: Lay Out the Sampling Locations on the Site

To locate the sample points on the site, use the same procedure as was used to construct the diagram of the sample points in Step 5, but use a tape measure or pacing, as appropriate, to measure distance. Since s = 1 in. in the diagram (Figure 10), then s = 4 ft on the site. Similarly, u = 3 ft 6 in. on the site. It may be helpful to show the actual distances (in ft) on the diagram before laying out the site sample points. For example, the samples on the wall are most easily found by measuring the distance on the scaled diagram from one end of the wall and the height above the driveway, and then converting these measurements to find the actual location on the wall. Consider point 32, for example. On Figure 10, it is located approximately 3/4 in. above the driveway and 5/8 in. from the left edge of the wall. On the site, then, this point is 3 ft above the driveway and 2-1/2 ft from the left edge of the wall.

The PCB spill site should be considered contaminated until laboratory analyses of the samples taken verify the site is clean. Therefore, caution should be exercised when marking the sample points on the site to prevent possible cross-contamination. The inspector should make minimum contact with the spill surfaces. One method for accomplishing this would be to cover the surfaces with plastic sheeting.

G. Step 7: Consideration of Special Cases

1. Sample Points Outside the Spill Cleanup Area

Samples from points outside the spill area should generally be collected, although taking these samples is at the discretion of the inspector. Collection of these samples permits the EPA to check the contamination of samples outside the spill area designated by the party responsible for the cleanup. This provides a mechanism for assessing whether the spill area was underestimated by the cleanup crew.

In cases where the contaminated area is very different from a circle (e.g., a very elongated ellipse) the sampling circle may be a poor approximation of the contaminated area, and a moderate to large percentage of the sampling points may fall outside the contaminated area. If the sampler is certain that the spill boundaries truly represent the contaminated area (i.e., there is definitely no contamination outside of this area), then it is permissible to disregard those sampling points falling outside the contaminated area. However, it is still good practice to collect such samples because the effort required to return to the site and sample again (should these samples be needed for any reason) is much greater than the effort required while on site.

2. *Sample Locations Which Do Not Physically Exist*

The grid can also indicate sample locations which do not physically exist on the real site. These locations are in "thin air" so to speak and cannot be sampled. The number of samples to be collected is adjusted downward for these samples; replacement locations are not needed.

3. *Judgmental Samples*

The inspector's best judgment should be used to collect samples where residual PCB contamination is suspected. These samples would be collected in addition to those from the sampling grid. Examples of extra sampling points include suspicious stains outside the spill area, cracks or crevices, or any area where the inspector suspects inadequate cleanup.

4. *Sampling Small Areas*

The grid sample design specifies that seven samples should be taken in areas which have a sample circle radius of less than 4 ft. In cases where the spill area is very small, fewer than seven samples can be taken at the discretion of the EPA inspector.

H. Example of Laying Out the Sample Design

This section summarizes the step-wise procedures required to determine the locations of the grid sample points at a PCB spill site. The example used is a simple 8 x 10 ft rectangular spill site.

Steps 1 and 2: Measure and Diagram the PCB Spill Cleanup Site

The PCB spill cleanup site must first be measured (usually with a tape measure). Then the site should be drawn to scale on graph paper. In this example, the site is assumed to be an 8 x 10 ft rectangle, as shown in Figure 11. A scale of 1 in. = 2 ft is used.

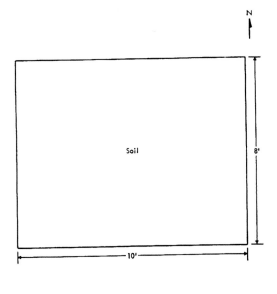

Figure 11. Scale diagram of PCB spill site.

Step 3: Determine the Center and Radius of the Sampling Circle

The center and radius of the sampling circle is determined on a separate diagram as follows, and is illustrated in Figure 12:

1. Draw the site diagram to scale (same as Figure 11).
2. Draw a line representing the longest dimension, L_1, of the site diagram.
3. Find the midpoint, P, of L_1.
4. Draw a second line, L_2, perpendicular to L_1, through point P. Line L2 must extend to the boundaries of the site.
5. Find the midpoint, C, of line L_2. Point C is the <u>center</u> of the sampling circle. (In this example, points P and C coincide, but will not coincide for many other types of configurations.)
6. Measure the distance from point C to either end of L_1, which is the sampling radius, r. The distance, r, should be measured to the nearest 1/16 in.
7. Scale radius, r, up to actual size. In this example, the radius, r, is 3-1/4 in. on a scale of 1 in. = 2 ft, or 6-1/2 ft (3-1/4 in. x 2 ft/in.).

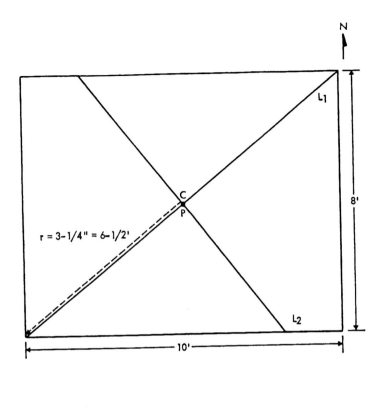

Figure 12. Determining center (C) and sampling radius (r) of sampling circle.

Step 4:   Find the Number of Grid Samples to be Used

The number of samples to be taken in a hexagonal grid depends upon the length of the sampling radius, as shown in Table 1 and repeated here.

Sampling radius, r (ft)	Number of Samples
≤ 4	7
> 4 - 11	19
> 11	37

Since the radius in this example is 6-1/2 ft, the number of sampling points would be 19.

Step 5: Plot the Sampling Points on the Site Diagram

The sampling points in a grid row are a distance, s, apart; and the grid rows are a distance, u, apart. The distances s and u are determined from the following table.

Number of Samples	Distance, s, Between Adjacent Sample Points	Distance, u, Between Successive Rows
7	0.87r	0.75r
19	0.48r	0.42r
37	0.30r	0.26r

In this example, the distance, s, between the points in a row is 1-9/16 in. [(0.48) x (3.25 in.)] on the diagram, or about 3 ft 2 in. [(1-9/16 in.) x (2 ft/in.)] on the actual site. The distance, u, between rows is 1-3/8 in. [(0.42) x (3.25 in.)] on the diagram, or about 2 ft 9 in. [(1-3/8 in.) x (2 ft/in.)] on the actual site.

The center point of the grid lies on the center, C, of the sampling circle. Construct the hexagonal grid and superimpose it over the site diagram (constructed on a third piece of graph paper), as illustrated in Figure 13 for this example. The middle row of the grid (points 1 through 5) should be oriented to maximize the number of sample points which lie within the boundaries of the spill cleanup site.

It should be noted that adjacent rows are staggered, and that the sample points of one row are located midway (horizontally) between the sample points of the other row.

Step 6: Mark the Sample Points on the Site

Starting at the center, C, of the spill cleanup site, mark the middle row points a distance of 3 ft 2 in. apart. Locate the adjacent rows a distance (u) of 2 ft 9 in. from the middle row, and mark the four sample points in each of these rows a distance of 3 ft 2 in. apart. Complete the site sampling grid with the other two rows of sample points.

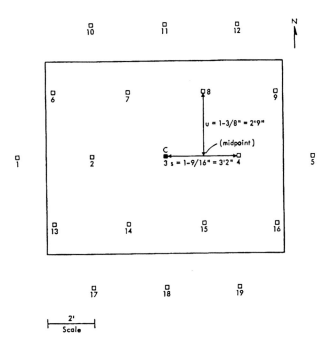

Figure 13. Diagram of 19-point grid superimposed on the PCB spill site.

VI. Sample Collection, Handling and Preservation

After the sampling grid has been diagrammed on the site description forms and laid out on the site, a sample must be taken at each grid point. Until the samples have been analyzed, the entire area must be assumed to be contaminated with residual PCBs. Therefore, appropriate measures must be taken to protect workers and the general public, prevent cross-contamination of samples, and prevent contamination of the surrounding area during sampling. Detailed contamination prevention procedures should be given in the staff training (Section III and Section VIII.B.).

PCB spill sites will vary widely in nature, and the types of media to be sampled may include soil, sod, water, hard surfaces, and vegetation. This section presents some general methods that can be used to sample these different media. These sample collection, handling and preservation techniques are provided for information; other techniques may also be used. Additional sample collection guidance documents are also available (Mason 1982; USEPA 1981).

A. Surface Soil Sampling

When surface soil (or sand) is to be sampled, the sample area should be marked by a 10 cm x 10 cm (100 cm^2) template. The soil should be scraped to a depth of about 1 cm with a stainless steel trowel, scoop, or spatula to yield about 100 g of soil. If more soil is required, the area should be expanded without increasing the depth of soil obtained. The soil sample should be placed in a precleaned glass bottle, the bottle capped, the sample bottle label filled out and attached, and a yellow TSCA PCB mark affixed. The bottle should be sealed in a plastic sample bag and placed in an ice chest containing ice (to keep the sample at about 4°C). If samples are to be analyzed soon, the cold storage requirements may be relaxed as long as sample integrity is maintained. The sample collection data should be entered in the field log book and on the chain-of-custody form.

The template used to mark surface soil samples, the scoop or spatula used to take the sample, and the rubber gloves worn by the inspector are all sources of cross-contamination between samples. Ideally, a different template, scoop, and pair of rubber gloves should be used to take each sample. The template and scoop may then be placed in a plastic bag to be taken back to the laboratory to be cleaned for the next field sampling job. The rubber gloves should be discarded into a plastic bag which will be disposed of as PCB contaminated material if any samples exhibit PCB contamination.

If a sufficient number of templates or scoops are not available to use only one item per sample, then each of these equipment items must be thoroughly cleaned between samples. The template and scoop should be thoroughly rinsed with solvent and wiped with a disposable wiping cloth (which should be discarded into the plastic bag intended for disposal of PCB contaminated materials).

B. Soil Core-Sampling

When core samples of sod or soil are needed, the samples may be taken using a coring device such as a piston corer or King-tube sampler. Core samples should be taken to a depth of about 5 cm. The soil core can be pushed out into a precleaned glass bottle and capped, or the tube containing the sample can be wrapped in solvent-rinsed aluminum foil, depending upon the type of coring device used. The sample should be properly labeled, a yellow TSCA PCB mark affixed, and placed in an ice chest (to keep the sample about 4°C). If samples are to be analyzed soon, the cold storage requirements may be relaxed as long as sample integrity is maintained. The sample collection data should be entered in the field log book and on the chain-of-custody form.

Core samples of soil or sod should be taken with individual core tubes for each sample. If this is not possible, then the coring device should be rinsed with solvent and wiped with a disposable wipe cloth to remove any visible particles before taking another sample. After each sample, rubber gloves and wipe cloth should be discarded into a plastic bag intended for disposal of PCB-contaminated materials.

C. Water Sampling

PCB spills on water may result in a surface film (particularly when the PCBs are dissolved in hydrocarbon oils) or sink to the bottom (particularly when the PCBs are in askarel or other heavier-than-water matrix). When a surface film is suspected (or visible), the water surface should be sampled. Otherwise, a water sample should be taken near the bottom of the body of water.

1. Surface Sampling

Surface water samples should be collected by lowering an open, precleaned glass sample bottle horizontally into the water at the designated sample collection point. As water begins to run into the bottle, slowly turn the bottle upright, keeping the lip just under the surface so that only surface water is collected. Lift the bottle out of the water, wipe the outside with a disposable wiping cloth, and cap the bottle. Label the bottle, affix a yellow TSCA PCB mark, and put the bottle in an ice chest (to keep the sample at about 4°C). If samples are to be analyzed soon, the cold storage requirements may be relaxed as long as sample integrity is maintained. The sample collection data should be entered in the field log book and on the chain-of custody form. The wiping cloth and rubber gloves should be discarded into a plastic bag used for disposal of PCB-contaminated materials.

2. Subsurface Sampling

Water near the bottom of the body of water should be sampled by lowering a sealed sampler bottle to the required depth, removing the bottle top, allowing the bottle to fill, and removing the bottle from the water. Transfer the subsurface sample into a precleaned glass bottle and cap. Wipe the bottle with a disposable wiping cloth, fill out and label the sample bottle, affix a yellow TSCA PCB mark, and put the sample bottle in an ice chest. If samples are to be analyzed soon, the cold storage requirements may be relaxed as long as sample integrity is maintained. The sample collection data should be entered into the field log book and on the chain-of-custody form. The wiping cloth and rubber gloves should be discarded into a plastic bag used for disposal of PCB-contaminated materials.

To prevent cross-contamination of samples, separate sampler bottles should be used to take the samples. Alternatively, the sampler bottle can be rinsed three times with distilled water, solvent-rinsed, and air-dried between samples.

Sometimes the above approaches to water sampling are not feasible. In these cases, other equipment such as siphons, pumps, dippers, tubes, etc., may be used to collect a water sample and transfer it to a precleaned glass sample bottle. The sampling system should be constructed of glass, stainless steel, Teflon, or other inert, impervious, and noncontaminated materials. Water samples taken with siphons, dippers, tubes, pumps, etc., may become cross-contaminated if the equipment is not cleaned between samples. Equipment cleaning may be achieved in most cases by flushing the equipment with distilled water and solvent.

D. Surface Sampling

Samples of hard surfaces may be taken by two methods: (a) wipe sampling and (b) destructive sampling. Wipe samples are taken of any smooth surface which is relatively nonporous (such as rain gutters, automobiles, and aluminum siding), while destructive samples are taken of hard porous surfaces (such as concrete, brick, asphalt, and wood). Both wipe and destructive samples may be taken if it is not known whether the surface is porous or not.

1. Wipe Sampling

A wipe sample is taken by first applying a suitable solvent (such as isooctane) to a piece of 11 cm filter paper (e.g., Whatman 40 ashless or Whatman 50 smear tabs) or gauze pad. The moistened filter paper or gauze pad is then held with a pair of stainless steel forceps or rubber gloves and rubbed thoroughly over a 100-cm^2 area (delineated by a template) of the sample surface to obtain the sample. The filter or pad is placed in a precleaned sample bottle, which is then capped, labeled, affixed with a yellow TSCA PCB mark, and placed in an ice chest (to keep the sample at about 4°C). If samples are to be analyzed soon, the cold storage requirements may be relaxed as long as sample integrity is maintained. The sample collection data are entered into the field log book and on the chain-of-custody form.

The template should be thoroughly rinsed with solvent and wiped with a disposable wiping cloth. The rubber gloves worn when taking wipe samples and the wiping cloth should be discarded into a plastic bag for disposal of PCB-contaminated materials.

2. Destructive Sampling

Wipe sampling is not appropriate on some porous surfaces, such as wood, asphalt, concrete, and brick, which will absorb the PCBs. In some cases, these surfaces can be sampled by taking a discrete sample such as a piece of wood or paving brick. Otherwise, chisels, drills, hole saws, etc., can be used to remove sufficient sample for analysis. Samples less than 1 cm deep should be taken and placed in a glass sample bottle or solvent-rinsed aluminum foil. Each sample container should be labeled, affixed with a yellow TSCA PCB mark, and placed in an ice chest. If samples are to be analyzed soon, the cold storage requirements may be relaxed as long as sample integrity is maintained. Sample collection data should be entered into the field log book and on the chain-of-custody form.

Equipment used to take samples of wood, asphalt, etc., should be cleaned with solvent and wiped between samples. Also, rubber gloves and wipe cloths should be discarded into a plastic disposal bag intended for PCB contaminated materials.

E. Vegetation Sampling

The sample design or visual observation may indicate that samples of vegetation, such as tree leaves, bushes, and flowers, are required. In this case, the sample may be taken with pruning shears, a saw, or other suitable tool, and placed in a precleaned glass bottle, which should be capped, labeled, affixed with a yellow TSCA PCB mark, and placed in an ice chest. If samples are to be analyzed soon, the cold storage requirements may be relaxed as long as sample integrity is maintained. The sample collection data should be entered into the field log book and on the chain-of-custody form.

After each sample is taken, the pruning shears should be rinsed with solvent and wiped with a disposable wipe cloth to prevent cross-contamination between samples. Also, rubber gloves and wipe cloths should be discarded into a plastic disposal bag intended for PCB-contaminated materials.

F. Compositing Strategies

Compositing is the pooling of several samples to form one sample for chemical analysis. In many circumstances it may be desirable to composite samples to reduce the number of (often costly) analyses needed. The suggested strategies for compositing samples are given in the appendix.

VII. Quality Assurance

Quality assurance must be applied throughout the entire sampling program, including sample design and sample collection, handling, and preservation. Each EPA office must develop a quality assurance plan (QAP) according to

EPA guidelines (USEPA 1980). The QAP must be submitted to the regional QA officer or other appropriate QA official for approval prior to sampling PCB spill sites.

The elements of a QAP (USEPA 1980) include:

- Title page
- Table of contents
- Project description
- Project organization and responsibility
- QA objectives for measurement data in terms of precision, accuracy, completeness, representativeness, and comparability
- Sampling procedures
- Sample tracking and traceability
- Calibration procedures and frequency
- Analytical procedures
- Data reduction, validation, and reporting
- Internal quality control checks
- Performance and system audits
- Preventive maintenance
- Specific routine procedures used to assess data precision, accuracy, and completeness
- Corrective action
- Quality assurance reports to management

Each EPA inspector who will sample PCB spill sites should understand and conform with all elements of the QAP.

VIII. Quality Control

Each EPA office that samples PCB spill sites must operate a formal quality control (QC) program. The minimum requirements of this program consist of preparing field blanks for the laboratory; sampling without contamination of samples; maintaining a rigid chain-of-custody procedure for the samples; and fully documenting the entire sampling program and maintaining records of the documentation.

The quality control measures taken by each EPA office should be stipulated in the QA plan. The QC measures discussed below are given as examples only. EPA offices must decide which of the following measures, and additional measures, will be required for each situation.

A. Field Blanks

Field blanks are given to the laboratory to demonstrate that the sampling equipment has not been contaminated. A field blank may be generated by using the sampling equipment to obtain a clean sample of solids or water. For example, the scoop or soil coring device can be used to obtain a clean solids blank sample. The water sampling equipment can be used to collect a blank sample using laboratory reagent grade water. These field blanks should be obtained both before and after field sampling.

Field blanks for wipe samples should be obtained in the field by wetting a clean filter paper with the solvent and storing the wetted paper in a clean sample jar.

One empty glass sample bottle and one filled with solvent should also be given to the laboratory as field blanks.

B. Sampling Without Contamination

Samples collected from PCB spill sites which have been cleaned up may become contaminated in two ways: (a) dirty sample containers, and (b) cross-contamination of samples from the use of contaminated sampling equipment. The first type of contamination can be eliminated by properly precleaning all sample containers prior to making the sampling trip. All glass jars should be washed with soap and water, rinsed three times with distilled water, rinsed with solvent (isooctane is recommended), baked in an oven at 350°C for 1 h, and sealed with a Teflon-lined cap. All aluminum foil used should be rinsed with solvent.

The sampling equipment should be precleaned before the site visit by rinsing with solvent and thoroughly wiping the equipment down. Crosscontamination during sampling can be avoided by using a separate sampler (such as a scoop, spatula, corer, etc.) for each sample, or cleaning the sample equipment between samples. Methods that can be used to clean the equipment between samples are given in the sample collection, handling, and preservation discussion (Section VI).

C. Sample Custody

As part of the quality assurance plan, the chain-of-custody protocol must be described. A chain-of-custody provides defensible proof of the sample, and data integrity. The less rigorous sample traceability documentation merely provides a record of when operations were performed, and by whom. Sample traceability is not acceptable for enforcement activities.

Chain-of-custody is required for analyses which may result in legal proceedings, and when the data must be subject to legal scrutiny. Chain-of-custody provides conclusive written proof that samples are taken, transferred, prepared, and analyzed in an unbroken line as a means to maintain sample integrity. A sample is in custody if:

- It is in the possession of an authorized individual.
- It is in the field of vision of an authorized individual.
- It is in a designated secure area.
- It has been placed in a locked container by an authorized individual.

A typical chain-of-custody protocol contains the following elements:

1. Unique sample identification numbers.
2. Records of sample container preparation and integrity prior to sampling.
3. Records of the sample collection, such as:

   - Specific location of sampling.
   - Date of collection.
   - Exact time of collection.
   - Type of sample taken (e.g., water, soil).
   - Initialing each entry.
   - Entering pertinent information on chain-of-custody record.
   - Maintaining the samples in one's possession or under lock and key.
   - Transporting or shipping the samples to the analytical laboratory.
   - Filling out the chain-of-custody records.
   - Chain-of-custody records accompanying the samples.

4. Unbroken custody during shipping. Complete shipping records must be retained; samples must be shipped in locked or sealed (evidence tape) containers. The addressee should be notified and prepared to receive the samples from the shipper.

D. Documentation of Field Sampling

In order to assure that the field sampling project has been thoroughly documented, the documents described in the next section should be used to maintain the quality of the project.

IX. Documentation and Records

Each EPA office is responsible for preparing and maintaining complete records of the field sampling operations. A detailed documentation plan should be prepared as a part of the QAP, and should be strictly followed. The following written records should be maintained for each field sampling operation:

- •Equipment preparation log book
- •Sample codes
- •Field log book
- •Site description forms
- •Chain-of-custody forms
- •Sample analysis request forms
- •Field trip report

A. Equipment Preparation Log Book

A log book should be maintained which lists the sampling equipment taken to each spill site. A detailed description of the cleaning and preparation procedures used for the sample collection equipment (templates, scoops, glass bottle, etc.) should be recorded.

B. Sample Codes

Each sample should be assigned a unique sample code and labeled accordingly when collected. The sample code should contain information on the site and which sampling point the sample represents. This sample code must be used to identify all sample records.

Each sample must also be labeled with a yellow TSCA PCB mark as described in 40 CFR 761.45 until it is determined to be PCB free.

C. Field Log Book

The EPA inspector should maintain a field log book which contains all information pertinent to the field sampling program. The notebook should be bound and entries be made in ink by the field inspector. All entries should be signed by the inspector.

At a minimum, the log book should include the following entries:

- •Owner of spill site
- •Location of spill site
- •Date(s) of sample collection
- •Exact times of sample collection

- Type of samples taken and sample identification numbers
- Number of samples taken
- Description of sampling methodology
- Field observations
- Name and address of field contact
- Cross-reference of sample identification numbers to grid sample points (shown on site description forms)

Since sampling situations will vary widely, no specific guidelines can be given as to the extent of information which should be entered into the field log book. Enough information should be recorded, however, so that someone can reconstruct the sampling program in the absence of the field inspector.

The field log book should be maintained in a secure place.

D. Site Description Forms

Serialized site description forms should be used to record the conditions of the site, provide sketches of the site, and show the location of the grid sampling points. The grid sampling points should be shown on dimensioned drawings and numbered. These forms should be accompanied by photographs (preferably Polaroid-type photographs) of the site. Each form and photograph should be signed and dated by the EPA inspector.

E. Chain-of-Custody Forms

Chain-of-custody forms should be completed and accompany the samples. These forms should contain the following information:

- Project site
- Sample identification number
- Date and time of sample collection
- Location of sample site
- Type of sample (soil, water, etc.)
- Signature of sample collector
- Signatures of those who relinquish and those who receive the samples, and date and time that samples change possession
- Inclusive dates of possession

F. Sample and Analysis Request Forms

A sample analysis request form should accompany the samples delivered to the laboratory. The field inspector should enter the following information on the form:

- •Project site
- •Name of sample collector
- •Sample identification numbers
- •Types of samples (soil, water, etc.)
- •Location of sample site for each sample
- •Analysis requested [analyte (i.e., total PCBs), method, desired method detection limit, etc.]
- •QC requirements (replicates, lab blanks, lab spikes, etc.) Special handling and storage requirements

The laboratory personnel receiving the samples should enter the following information on the form:

- •Name of person receiving the samples
- •Laboratory sample numbers
- •Date of sample receipt
- •Sample allocation
- •Analyses to be performed

G. Field Trip Report

The EPA inspector should prepare a brief field trip report to be maintained on file. The report should provide information such as the project site, date(s) of sampling, types and number of samples collected, any problems encountered, any notable events, and specific reference to the other documents listed above.

X. Validation of the Manual

A previous draft of this manual entitled "Field Manual for Verification of PCB Spill Cleanup" (Draft Interim Report No. 3, Task 37, EPA Prime Contract No. 68-02-3938, June 27, 1985) was used in a brief field validation study. The primary purposes of the study were to: (1) determine the degree of difficulty of understanding the grid sampling designs in the field manual; (2) determine the amount of time and degree of difficulty required to lay out the sampling grids on simulated PCB spill sites; and (3) identify any concerns or problems that may arise in implementing the field manual. To achieve these goals, simulated PCB spill sites were constructed for the exercise. Four persons (Mr. David Phillippi and Mr. Robert Jackson of the EPA Region VII Office and Ms. Joan Westbrook and Mr. Ted Harrison of MRI) were selected to lay out the sampling grids on the spill sites after they had read the field manual. These four persons had no prior association with developing the field manual. Other persons from EPA and MRI acted as observers since they were intimately familiar with the field manual.

Four simulated spill sites having the following characteristics were laid out:

- A rectangle (3 ft x 6 ft)
- A parallelogram (about 3 ft on a side)
- A circle (about 12 ft diameter)
- A square (6 ft on a side)

The first two sites required seven grid sample points, and the other two required 19 grid sample points.

Each of the four "inspectors" laid out the grid sample points on two of the four sites after constructing the designs on graph paper. In all cases the sample points were laid out correctly with little or no difficulty in 30 min or less. Each inspector commented that there was little or no difficulty in performing the exercises.

As a final exercise, a large irregular simulated PCB spill site was constructed, and all attendees participated in laying out the 37 grid sample points. The spill site was designed so that some sample points were located on the floor and two adjacent walls to make the exercise relatively difficult. The 37 grid sample points were laid out correctly with relative ease in about 45 min. Some discussions were required to decide how to treat sampling points which fell in the overlap where the two walls intersected.

It was concluded from the exercise and discussions which followed that: (1) the field manual is easy to follow and understood by people unfamiliar with the manual prior to reading it; (2) the grid sample points are never "perfectly" laid out (with the sample points precisely aligned) so that some degree of randomness is built into the sample designs; (3) the time required to lay out the grid sample points after the boundaries of the spill site have been determined is relatively short (less than 1 h); and (4) using this manual, the grid sample points can be correctly laid out by inexperienced people.

XI. References

Boomer BA, Erickson MD, Swanson SA, Kelso GL, Cox DC, Schultz BD. 1985 (August). Verification of PCB spill cleanup by sampling and analysis (second printing). Interim report. Washington, DC: Office of Toxic Substances, U.S. Environmental Protection Agency. EPA-560/5-85-026.

Mason BJ. 1982 (October). Preparation of soil sampling protocol: techniques and strategies. ETHURA, McLean, VA, under subcontract to Environmental Research Center, University of Nevada, for U.S. Environmental Protection Agency, Las Vegas.

USEPA. 1980. U.S. Environmental Protection Agency. Guidelines and specifications for preparing quality assurance project plans. Office of Monitoring Systems and Quality Assurance, QAMS-005/80.

USEPA. 1981 (March). U.S. Environmental Protection Agency. TSCA Inspection Manual.

Strategies for Compositing Samples

Appendix

This appendix gives suggested strategies for compositing samples taken from PCB spill sites which are sampled using the grid sampling methods described in the text of the report. Compositing may result in a savings of analysis time and cost. Sample compositing is not required and should be used only if time or cost savings may result. The strategies for forming composites are as follows:

1. Composite only samples of the <u>same type</u> (i.e., all soil or all water). Since the composite must be thoroughly mixed to ensure homogeneity, certain types of samples such as asphalt, wipe samples, wood samples and other hard-to-mix matrices should not be composited.

2. Do not form a composite with more than 10 samples, since in some situations compositing a greater number of samples may lead to such low PCB levels in the composite that the recommended analytical method approaches its limit of detection and becomes less reliable.

3. For each type of sample, determine the number of composites to be formed using the table below.

Number of samples	Number of composites
2-10	1
11-20	2
21-30	3
31-37	4

As much as possible, try to form composites of equal size. For example, if 37 soil samples are taken, then four composites could be formed using 9, 9, 9, and 10 samples apiece.

4. To the extent possible, composite adjacent samples. If residual contamination is present, it is likely that high PCB levels will be found in some samples taken close together.

Because of the large number of situations that may be encountered in practice, it is not possible to specify compositing strategies more precisely. The laboratory and field staff should exercise judgment in all cases.

Part III

Cleanup
of PCB Spills

Bruce A. Boomer, Mitchell D. Erickson and Gary L. Kelso

I. Introduction

The U.S. Environmental Protection Agency (EPA), under the authority of the Toxic Substances Control Act (TSCA) Section 6(e) and 40 CFR Section 761.60(d), has determined that polychlorinated biphenyl (PCB) spills must be controlled and cleaned up. The Office of Toxic Substances (OTS) and Midwest Research Institute have studied the cleanup, sampling and analysis, and verification of cleanup of PCB spills and have issued a series of reports.

A previous report on this work assignment, "Verification of PCB Spill Cleanup by Sampling and Analysis" (Boomer et al. 1985), outlined specific sampling and analysis methods to determine compliance with EPA policy on the cleanup of PCB spills. A second report, "Field Manual for Grid Sampling of PCB Spill Sites to Verify Cleanup" (Kelso et al. 1986) provided detailed, step-by-step guidance to EPA enforcement staff in determining whether a PCB spill site has been cleaned up to the EPA-mandated level of cleanness.

This report provides general information on how to clean up a PCB spill. Although the primary audience is any user organization not already familiar with PCB spill cleanup procedures, the information presented in this report provides useful suggestions for any organization involved in PCB cleanup activities.

This report is not a detailed manual. The approach and methods for cleaning up a PCB spill are suggestions only; other methods may be equally effective depending on the circumstances. Although this report addresses the cleanup of PCB spills primarily from pole-mounted capacitors, the information presented also provides a general background for cleanup of PCB spills involving larger quantities of PCB-containing fluids.

0-87371-945-X/93/0.00 + $.50

This report is not a statement of policy. The information contained in this report does not reflect EPA policy except where specifically accompanied by a regulatory reference. Future changes in EPA policy may affect some of the information presented in this document.

Important sources of information for this document included discussions with experienced staff from EPA offices, electric utilities, and industrial organizations. Additional insight was provided by PCB experts at a working conference on PCB spills sponsored by EPA-OTS and held at MRI in February 1985.

Following a summary of the report (Section II), Section III outlines regulations and policy for cleaning up PCB spills and describes the types of spills and spill scenarios. Section IV discusses the cleanup process, emphasizing health and environmental protection measures and presents a general approach to cleaning up a PCB spill. Optional activities are addressed in Section V.

II. Summary

This report provides general information on how to clean up a polychlarinated biphenyl (PCB) spill. The U.S. Environmental Protection Agency (EPA), under the authority of the Toxic Substances Control Act Section 6(e) and 40 CFR Section 761.60(d), has determined that PCB spills must be controlled and cleaned up. This report is neither a detailed manual nor a statement of policy. The described approach and methods are suggestions based on the experience of staff from EPA offices, electrical utilities, and industrial organizations.

The EPA considers any spill, leak, or other uncontrolled discharge of PCBs at concentrations of 50 ppm or greater to be improper disposal of PCBs. The spills can be generally characterized in terms of the type of failed equipment causing the spill, the human/environmental exposure scenario, the volume/concentration of spilled PCBs, the area of the spill, and the materials contaminated by the spill.

Health and environmental protection are important concerns in the cleanup process. Throughout the cleanup, major points of emphasis include protecting public health, protecting the health and safety of the cleanup crew, keeping PCBs out of sewers and waterways, and avoiding further dispersion or cross contamination of PCBs.

Cleaning the PCB spill site may include:

1. Reporting the spill.
2. Quick response and securing the site.
3. Formulating a cleanup plan.
4. Cleanup.
   - Removal of failed equipment
   - Decontamination of pavement
   - Removal of sod and soil
   - Removal of contaminated vegetation
   - Removal/decontamination of other materials
5. Proper disposal of contaminated materials.
6. Post-cleanup activities.
   - Sampling and analysis
   - Remedial action
   - Returning the site to normal use

Some optional activities may minimize future concerns associated with PCB spills, prepare for efficient response to future spills, and improve public relations associated with PCB spill incidents. These activities may include additional sampling and analysis, record keeping, staff training, maintaining PCB spill cleanup kits, and taking steps to maintain good public relations.

III. Overview of PCB Spills

This section outlines the regulatory background and concerns associated with PCB spills and also describes several types of spill scenarios. Individuals responsible for planning or implementing PCB spill cleanup should be aware of the requirements, concerns, and decisions associated with PCB spills.

A. Regulations and Policy for PCB Spills

The EPA, under the authority of the Toxic Substances Control Act Section 6(e) and 40 CFR Section 761.60(d), has determined that PCB spills must be controlled and cleaned up. EPA regulations controlling the disposal of PCBs, promulgated in the Federal Register on February 17, 1978 (43 FR 7150) and May 31, 1979 (44 FR 31514), broadly define the term "disposal" to encompass accidental as well as intentional releases of PCBs to the environment. Under these regulations, EPS considers spills, leaks, and other uncontrolled discharges of PCBs at concentrations of 50 parts per million (ppm) or greater (whether intentional or unintentional), improper disposal of PCBs. For such spills, EPA has the authority under Section 17 of TSCA to compel persons to take actions to rectify damage and cleanup contamination resulting from the spill.

Since 1978, EPA standards for the cleanup of spilled PCBs have been established in the EPA Regional Offices in the form of general guidelines applied on a case-by-case basis for specific spill situations.

B. Policy and Definitions for This Report

This report does not address specific EPA standards for the cleanup of spilled PCBs.   Rather, this report gives a general approach which is in accordance with recent practice.   Future changes in policy may add specific EPA guidance or standards to the general suggestions presented in this report. Details on current EPA national or regional standards for the cleanup of spilled PCBs can be obtained by calling any EPA Regional Office.

Due to the general nature of this report, it is necessary to define the following two terms as they are used in this publication:

- •PCBs - Polychlorinated biphenyls as defined in 40 CFR 761.3.
- •PCB spill - Any spill, leak, or other uncontrolled discharge of PCB at concentrations of 50 ppm or greater (whether intentional or unintentional). The concentration of PCBs spilled is determined by the PCB concentration in the original material spilled.

C. Types of PCB Spills

PCB spills generally are viewed as unique situations to be evaluated on a case-by-case basis in terms of cleanup methods and EPA enforcement.   Spills can be categorized roughly, however, in terms of the type of failed equipment causing the spill, the human and environmental exposure scenario, and other contributing factors (volume or concentration of spilled PCB materials, area of spill, and materials contaminated by the spill).

1. Types of Failed Equipment Causing PCB Spills

Most PCB spills involve failed electrical equipment, although spills from the transportation or other handling of PCBs can also occur.   Spills of PCBs in EPA-approved storage areas (40 CFR 761.65) will be contained by curbed containment areas; these spills are some of the least demanding PCB spill situations since the extent of the spill is known.

A large proportion of the reported PCB spills are associated with the distribution, substation, and network electrical equipment used by electrical utilities and related organizations.   The types of equipment and PCB spill incidents typically found within electric utilities are the following (Lincoln Electric System 1982):

* Distribution Capacitor

A spill involves a small amount of PCB (about 1 gal.), but PCB can be sprayed over a large area.

* Substation Capacitor

A spill involves a small amount of PCB (about 1 gal.), but PCB can be sprayed over other substation capacitors, concrete pads, ground, or other surfaces.

* Distribution Transformer

A spill may involve a large amount of oil containing PCBs. Although spills may be only slow leaks contained in a small area, oil can also be sprayed over a large area by wind or an explosion.

* Network Transformer

The network transformer may be filled with up to 70% PCB. These transformers are located in vaults, which provide some containment, but an explosion could spray PCB over a large public area. Ventilation is a special concern.

* Substation Equipment (In Addition to Capacitors)

A spill involves a large amount of oil (that may be contaminated with PCB) that has the potential to cover a large area.

Capacitor spills occur in the largest numbers. Edison Electric Institute reported that the electric utility industry owned 1.9 million PCB capacitors as of June 30, 1984, and that 0.06% of these were expected to leak or spill annually (USWAG 1984). (The average spill quantity was reported to be 1.5 gal.) Many electric utilities are replacing their PCB capacitors, however, with capacitors that do not contain PCBs.

Other PCB spill incidents may include slow leaks from various types of equipment and spills and leaks from equipment used in manufacturing and mining applications.

Fires associated with PCB equipment create special health and environmental problems, as discussed in the literature (USEPA 1985). PCB fires, however, are beyond the scope of this report.

2. Exposure and Dispersion Scenarios for PCB Spills

The potential health and environmental impact of a PCB spill is related to the potential for human exposure to PCB and to dispersion of the PCB, An assessment of potential exposure and dispersion of the spilled PCB is an important factor in selecting the level of effort and priorities of the cleanup activities.   Risk scenarios are grouped into the following approximate categories:

a. Highest Exposure Potential

Examples of the highest exposure potential include:

1. Spills that have the potential to result in the direct contamination of waterways, groundwater, or food and feed.
2. Spills resulting in indoor contamination.
3. Spills occurring in children's play areas (school yards, park playgrounds, or certain residential areas).

b. High Exposure Potential

The general mid-range category includes the following:

1. Spills in urban areas with potential human exposure (residential areas, public areas, and sidewalks and streets, etc.).
2. Spills in rural areas where the risks of human exposure are comparable to those in a residential situation.

c. Low Potential Exposures

This category includes these examples:

1. Spills in fenced substations.
2. Spills in rural areas with limited exposure potential.

Section IV of this report describes cleanup methods appropriate for categories "b" and "c" above. The cleanup of category "a" PCB spills will require a special level of caution and efficiency. Special advice from any EFA Regional Office or a state or local environmental agency may be appropriate.

3. Other Factors in Categorizing Spills

Individual company policies, regulatory requirements, and practical considerations for cleaning up a PCB spill are influenced by the volume and PCB concentration of the soiled material. Minimum reporting requirements, safety policy, and the appropriate level of effort for the spill cleanup are often related to the amount of PCB spilled. Similarly the area of the spill site and the amount and type of materials contaminated by the spill will set the practical requirements for the size of the cleanup crew, the type of equipment and supplies needed, and the general level of effort. The next section details the actual cleanup process.

IV. The Cleanup Process

Cleaning up a PCB spill involves a number of specific components and a constant awareness of the need to protect health and the environment. Following a discussion of the health and environmental issues, this section presents a general approach to PCB spill cleanup.

A. Health and Environmental Protection

PCBs are regulated because of the health-related concerns and potential environmental impacts associated with the improper storage, usage, or disposal of materials containing PCBs.   It is therefore essential that all persons participating in PCB spill cleanup activities take special precautions to protect the health of the public and the cleanup crew and to prevent the release of PCBs to the environment.   The following subsections provide suggested appropriate steps and precautions.

1. Public Health

Environmental and health agencies stress the importance of a quick response to a PCB spill in order to secure the site and prevent public exposure to the spilled materials. A study conducted by the Edison Electric Institute's Utility Solid Waste Activities Group concluded that restricting access to a spill greatly reduces overall exposure and that quick response reduces potential health effects (USWAG 1984). (Quick response and securing the spill site are discussed in more detail later in this section.)

2. Health and Safety of the Cleanup Crew

Protection of the health and safety of the cleanup crew during the PCB cleanup operation is an important concern. Although skin contact is the major worry, respiratory protection may also be necessary if ventilation is inadequate or if wind-borne dust could expose workers.

a. Skin Protection

In order to prevent skin contact with PCBs, the cleanup crew should wear protective clothing that is impervious to PCBs. Gloves, boots, aprons, or coveralls made of polyethylene, vinyl, or fluorelastomer are appropriate (Phillips 1983; Lincoln Electric System 1981).   The use of disposable overboots or multilayer protective gloves may provide additional cost-effective protection in some cases. Proper disposal of all protective clothing and any potentially contaminated personal clothing is preferable to cleaning or decontamination. If PCBs come into contact with the skin, wash with waterless soap and wipe with disposable towels. Then wash with warm water and soap, and properly dispose of cleaning materials with other contaminated wastes.

b. Eye Protection

Chemical safety goggles, face shields (8 in. minimum) with goggles, or safety glasses with side shields provide eye protection during PCB spill cleanup activities (NIOSH 1977). If PCBs come into contact with the eyes, flush the eyes immediately with large quantities of water and obtain professional medical assistance.

c. Respiratory Protection

In PCB spill cleanup the key issue in respiratory safety is ventilation; inhalation is reported to be the dominant exposure pathway for PCBs (USWAG 1984). PCB vapors, in general, however, are not considered to be a safety problem in open spaces unless large quantities of PCB or elevated temperatures are involved (Phillips 1983). However, wind-borne dust can cause respiratory safety concerns. Confined spaces should be force ventilated and respiratory protection may be appropriate.

For assistance in determining the appropriate level of respiratory protection for a specific incident or case, contact the regional office of the Occupational Safety and Health Administration or a similar state or local agency. References discussing health and safety considerations relevant to some PCB spill incidents include NIOSH Criteria for a Recommended Standard for Exposure to Polychlorinated Biphenyls (PCBs) (1977) and Health Hazards and Evaluation Report No. 80-85-745 (NIOSH 1980).

d. General Health and Safety

Common sense in personal hygiene (e.g., no eating or drinking on site) is needed to avoid ingestion or further contact with PCBs. General safety procedures including the use of protective equipment (such as hard hats and safety shoes and safe work practices will reduce the likelihood of injury (such as electrical shock and falls) to workers during the cleanup activities.

3. *Environmental Protection*

Throughout the PCB cleanup activities, the cleanup crew must take all steps necessary to minimize the release of PCBs to the environment. Key precautions include the following:

- Keep spilled PCBs out of sewers and waterways.
- Avoid cross contamination or dispersion or PCBs from workers' shoes, contaminated equipment, spilled solvents, rags, and other Sources.

B. Cleanup of the Spill

This subsection presents a general approach to cleaning up a PCB spill. The approach is based on the experience of PCB experts at EPA, electric utilities, and other organizations. The information does not reflect EPA policy except where noted specifically with a regulatory reference. The approach and methods described in the following pages are general suggestions; other methods may be equally effective depending on the circumstances. Often PCB spills will create unique problems requiring a case-by-case best judgement approach.

1. Reporting the Spill

As soon as it is detected, the spill must be reported to the PCB owner company or the authorities by the employee or individual who discovers the spill. Within the owner company, the responsible staff must report the spill to federal, state, and local authorities (as required) and make arrangements to secure the site. Based on the type of equipment involved in the spill, records of the PCB content of the equipment, and the nature of the spill, the company must quickly determine if the spill exceeds the regulatory limits. When in doubt, it is best to report the spill.

Under EPA regulations [50 FR 13456-13475], spills of over 10 lb of PCB must be reported to the National Response Center. The toll-free phone number is (800) 424-8802. State and local regulations may require additional spill reports.

Spills of 10 lb or less of PCB may create a special concern if the spill occurs in an area where there is significantly high potential human exposure (e.g., indoors, school yards, park playgrounds) or into a waterway or drinking water source. Such spills should be reported to the local authorities so as to determine the best methods to minimize the health and environmental impact of the spill.

When appropriate, property owners and residents should be notified as soon as possible after a PCB spill has occurred on their property or in their immediate vicinity. Public safety and cooperation are important goals.

2. Quick-Response Containing and Securing of the Site

a. Response

Quick response is desirable in order to reduce the dispersion of the spilled material and to prevent access to the site. Federal regulations require that cleanup actions for leaking PCB transformers commence within 48 hours of discovery of a spill [40 CFR 761.30(a)(1) (iii)]. A more rapid response is highly preferable for any PCB spill incident.

A quick response allows containment of the spill and removal or cleaning up of the PCB-contaminated material before it is dispersed by wind, rain, seepage, or other natural cause, or by humans or animals. Also, a smaller spill area will typically cost less to clean up than a spill area that has "grown" due to dispersion of PCB. Since PCB spills from pole-mounted electrical equipment often occur during storms or other extreme weather conditions, response may need to be delayed. Nevertheless, the spill site should be contained and secured as soon as possible.

b. Containing the Spill

The first priority is to contain the spill while taking appropriate health and safety precautions. Containing the spill may require stopping the flow from the source (if still leaking) and preventing the spread of the spilled material.

Leaks may be stopped by the following methods (USEPA 1982):

- Turning off spigots or control valves.
- Plugging the leak with rags, plugs, or other materials from the inside or outside.
- Patching the leak with an appropriate patching compound.
- Placing the leaking article into a container or diked area.
- Changing the position of the container so that the leak hole is at the highest level.
- Collecting the spillage in a container located under the leak, channeling the leak, or pumping the leaking contents into a container or storage area.

Use dams, dikes, or trenches if necessary to control the spread of the spilled material, confining the spill to the smallest area possible.

c. Securing the Site

In securing the site two concurrent activities take place: (1) preventing access to the spill site, and (2) determining the approximate boundary of the spill. The site should be isolated by barricades, ropes, plastic cover, or other appropriate measure in order to prevent access or contamination spread by the public, vehicles, animals, or water and weather. Warning signs, blinking lights, or continuous surveillance at the site may be appropriate to prevent inadvertent public access.

To isolate the site effectively, the extent of the spill must be estimated by looking for stains or droplets of the PCB liquid and adding a buffer area that may include PCBs finely dispersed from splattering. Droplets may be noted only at relatively fresh spills but stains will probably persist for some time. USWAG (1984) also suggests noting the wind direction and orientation of stains around the pole (if applicable) in addition to looking for pieces of the affected equipment and damaged vegetation as part of the site evaluation.

Site inspection should be conducted by only one person, and that inspector should take precautions to avoid tracking PCBs from contaminated areas into clean ones. The inspector should plan out the area to be inspected, put on protective boots before entering the buffer zone of the contaminated area, and walk carefully from the perimeter inward, removing the contaminated boots immediately upon leaving the contaminated area (USWAG 1984).

Field analysis kits may aid in determining the extent of the spill in some instances. Field analysis kits, when used properly, can serve as a screening tool. Gauger (1985) provided data to support the use of a PCB field test kit to screen soil samples for PCB. The kit, which uses a chloridespecific ion electrode, appeared to have difficulties in detecting PCBs in wet soils but was shown to detect PCB below 20 ppm in various air-dried soils.

The need for quick response has limited the usefulness of the more accurate field analytical techniques such as gas chromatography. Practical problems associated with availability of the equipment and trained staff, setup time, and cost have so far limited the use of such techniques. Another analysis option, sending preliminary samples to a laboratory to determine the extent of a spill, could cause a significant delay in the cleanup process. A quick cleanup of the visibly contaminated area, followed by post-cleanup verification sampling (if appropriate), is usually the most effective cleanup method.

Selection of a buffer zone around the spill site will be a best judgment decision. For a capacitor spill, USWAG (1984) suggests the selection of an additional buffer zone area equivalent to one-half of the visible spill area. The buffer zone should completely surround the spill site. Detroit Edison Company surrounds each visible spill area with a buffer zone at least 5 feet wide (Eisele 1985).

After isolating the spill site, the contaminated area should be covered with secured plastic sheeting if cleanup activities do not begin immediately. The sheeting ultimately will be disposed of with the other contaminated materials resulting from the cleanup.

3. Cleanup Plan

After securing the site, the cleanup crew will either (a) proceed immediately with the cleanup operation, or (b) identify the materials spilled and formulate an appropriate cleanup plan. In addition to evaluating the site-specific factors (i.e., area of the spill site, potential for human exposure and environmental risk), the plan can be developed by estimating the concentration of PCBs in the spilled liquid, the volume of liquid spilled, and the total amount of PCBs spilled. By evaluating these basic factors, the individuals directing the cleanup can determine the necessary level of effort in accordance with the policy of the PCB owner and the EPA Regional Office (or other authority). The cleanup leader can determine if additional guidance is needed, plan sampling and analysis as appropriate, and make other decisions related to the level of effort and procedures needed.

4. Cleanup Procedures

Typically cleanup will involve some combination of the following steps, keeping in mind the health and environment-related objectives of the cleanup. In practice, cleanup should begin within 24 hours or as soon as possible.

Before removing contaminated material from the spill site, it is advisable to cover the area immediately surrounding the spill site (i.e., the areas bordering the "buffer zone") with heavyweight (about 10 mil) polyethylene sheeting secured with weighted objects to protect the surrounding area from contamination that results from spillage or tracking. Such sheeting should also be placed wherever drums filled with PCB waste are stored.

Cleaning or decontamination with an appropriate solvent or cleaning agent is suggested for specified situations. Cleanup crews may select materials previously used by various organizations for spill cleanup. These materials consist of kerosene, hexane, methylene chloride, fuel oil, and various proprietary solvents, detergents, and industrial cleaning agents. Selection will be based on the wettability of the surfaces, potential surface damage, and the ability of the solvent or cleaner to remove the PCBs from the contaminated surface.

a. Removal of Failed PCB Equipment

One of the first steps is to remove the failed equipment and any scattered equipment parts from the spill area. Appropriate safety precautions should be used. For pole-mounted equipment spills, equipment removal is followed by cleaning the pole and any other equipment or wiring in the immediate spill area (including the equipment of other utilities if applicable). Wipe down the items three times with an appropriate solvent (such as kerosene); use as little solvent as practical on the cleaning rag since splashing or liberal use of the solvent may increase subsurface migration (USWAG 1984). For all cleaning activities, clean from the outside of the spill site toward the inside (that is, from areas of lower suspected levels of contamination toward areas of higher suspected contamination). Cleaning rags should be changed often. Absorbent material (i.e., sawdust or "floor-dry" type material) placed at the base of the pole can collect contaminated solvent. Use a level of caution appropriate for the selected solvent. All contaminated materials must be collected for proper disposal or decontamination.

b. Contaminated Pavement

All free-flowing material should be removed with absorbents such as sawdust or "floor dry." After removing the absorbent material, the surface should be wiped down three times with rags or cloths soaked with an appropriate solvent or industrial cleaning agent. Additional cleaning could include soaking the pavement again with the solvent or cleaning agent and absorbing the residue with sawdust or "floor dry" (USWAG 1984). Never hose down PCBs with water.

Cracks and joints in the pavement may collect PCBs or allow seepage of PCBs into the ground beneath the pavement. Additional rinses, wipes, or other special efforts may be necessary to clean the cracks and joints effectively. Contamination of soil or other material below the crack or joint may require removal in some cases.

Some spills on pavement surfaces may pose special concerns of residual contamination and high potential for human exposure to PCBs. Such cases may necessitate removal of the contaminated pavement or encapsulation of the contaminated area under a coating impervious to PCBs, subject to EPA approval.

c. Removal of Sod and Soil

PCB spills onto fields and lawns are cleaned by removing the stained area and buffer zone of vegetation and soil, typically with a shovel, and placing the contaminated materials into drums for disposal. Contaminated sod and soil should be removed by working from the perimeter (buffer zone) toward the center of the spill. This technique is an attempt to minimize unintentional cross contamination. Manual removal with a shovel may be the most efficient and practical method of removing contaminated sod and soil, even for spills with larger areas. Bulldozers, backhoes, and similar equipment are usually not used because of the cross-contamination problems that they tend to create.

Kansas Electric Cooperatives suggests removal of both lawn grass and at least 2 in. of underlying soil. It also recommends that good judgment be used to determine the depth of PCB penetration during the period since initial contamination (Phillips 1983). USWAG (1984) suggests that after removing the sod the crew should employ a skimming method for soil removal. A skimming technique is preferred over a digging technique since PCBs do not normally percolate or soak down through most soils. The crew should skim the soil by 1-in. layers, changing their gloves and boots after removing each layer.

The company procedures of Detroit Edison Company include excavation of soils at least 12 in. deep (Eisele 1985).

Spills on sand, gravel, shells, or other porous media require a greater removal depth. Also, a standing puddle of PCBs (such as at the base of a pole) may flow downward, requiring deeper digging.

Cleaned areas should be covered with new plastic sheeting to minimize recontamination by the cleanup crew or the public (USWAG 1984). A plastic cover is particularly useful when the cleanup requires more than one day of labor or when the results of sampling and analysis are required before the site can be restored or closed.

Regardless of the method selected, the primary factors are removing the contaminated sod and soil, depositing the material in drums suitable for proper disposal, and avoiding further dispersion of PCBs by cross contamination.

d. Removal of Large Contaminated Vegetation

Brush, bushes, tree limbs, and other larger vegetation showing stains or other physical signs of PCBs should be trimmed back beyond the signs of contamination. Contaminated materials should be placed in appropriate containers for disposal.

e. Other Contaminated Surfaces or Objects

Good judgment is needed in deciding which contaminated items should be replaced or which items are appropriate for decontamination. Items of lower cost may be easier to replace. However, all items posing a potentially high exposure risk will require special consideration. The current decontamination requirements are subject to EPA approval on a case-by-case basis. When in doubt as to an appropriate level of cleanup activity, contact EPA Regional Office or state agency staff for advice.

f. Spills Into Water

Spills into waterways or potential drinking water sources create special environmental problems. All spills into a waterway, sewer, or any body of water should be reported to the National Response Center [phone (800) 424-8802] and to the EPA and state and local authorities, who will be able to provide directions for an appropriate cleanup response. For PCB spills into water, preventive action must be taken immediately to prevent additional dispersion of PCBs.

g. Summary

A simple PCB spill cleanup may involve removing of the leaking equipment, cleaning contaminated pavement with an absorbent material and solvents, removing contaminated sod and soil by shovel, and decontaminating or disposing of the equipment (shovels, shoes, gloves, rags, plastic sheets, etc.). More complicated situations may include removal of additional contaminated items and decontamination of fences, buildings, and electrical equipment.

At this point, however, the cleanup is still not complete. The remaining steps consist of proper disposal of all contaminated materials and completion of post-cleanup activities, which may include verification of adequacy of the cleanup by sampling and analysis, remedial action, site restoration, and other practical matters. These activities are presented in the following sections.

5. Proper Disposal of PCB-Contaminated Materials

All PCB-contaminated materials removed from the spill site must be shipped and disposed of in accordance with relevant federal, state, and local regulations. TSCA regulations [40 CFR 761.60] outline the requirements for the disposal of PCBs, PCB articles, and PCB containers in an incinerator, high efficiency boiler, chemical waste landfill, or an approved alternative method. Facility requirements for incineration and chemical waste landfills are presented in 40 CFR 761.70 and 40 CFR 761.75, respectively. Applicable U.S. Department of Transportation regulations are given in 49 CFR 172.101.

6. Post-Cleanup Activities

After the contaminated material has been decontaminated or removed for proper disposal, other issues may need to be addressed. Depending on the specifics of the spill, the EPA or other authority may require demonstration that the cleanup was adequate based on the results of sampling and analysis of the spill site. Suggested methods for verification by EPA enforcement staff of PCB spill cleanup by sampling and analysis were presented in two previous EPA reports (Boomer et al. 1985; Kelso et al. 1986).

If the results of any required analysis indicate that the cleanup was not in compliance with designated cleanup levels, additional cleanup will be needed. Additional sampling can pinpoint the location of remaining contaminated areas if the original sampling plan was not designed to identify contaminated sub-areas within the spill site. If additional cleanup is needed, the cleanup crew will continue as before, removing more material or cleaning surfaces more thoroughly. This remedial action will be followed by additional sampling and analysis to verify the adequacy of the cleanup.

As a practical matter, security of the site should be maintained (including a plastic sheeting cover over the "cleaned" area if appropriate) until any necessary cleanup standards are verified. This will minimize the potential dispersion of PCBs or cross contamination of the site.

After receiving any necessary approvals, the crew may complete the cleanup operation by restoring the site if applicable. Site restoration, which is not addressed under TSCA, is a matter to be settled between the company responsible for the PCB spill, the property owner, and other involved parties. Depending on the site, restoration may or may not be appropriate. However, special situations may suggest the placement of a soil cap or the use of encapsulating media on portions of the site, depending on EPA judgment.

Any potentially contaminated equipment, personal items, plastic sheeting, etc., must be removed for proper disposal before the cleanup crew returns the site to normal use.

V. Optional Activities

In addition to the basic elements addressed in the preceding section, some optional activities may minimize future concerns associated with past spills, prepare for efficient response to future spills, and improve public relations associated with PCB spill incidents.

A. Sampling and Analysis

Even if it is not required for a particular PCB spill incident, sampling and analysis can be used to determine whether the spilled material contains PCBs, to assess the degree of PCB contamination at a site prior to cleanup, or to verify the adequacy of the decontamination after a preliminary or final cleanup. The results of analysis can be used to verify compliance with a cleanup standard, to determine if additional cleanup is necessary or appropriate, or to indicate any special needs or precautions appropriate for future usage or contact with the spill site and its contents.

The results of the analysis may be useful documentation for any future inquiries related to the spill or for limiting current or future public concern associated with the spill. Sampling and analysis methods appropriate for PCB spill verification were presented in two EPA reports (Boomer et al. 1985, Kelso et al. 1986).

B. Records

Although there are no current TSCA requirements for maintaining records of PCB cleanup activities except for documentation of PCBs stored or transported for disposal [40 CFR 761.80(a)], the PCB owner should keep records of each spill cleanup in case of future questions or concern. Relevant information may include a chronology of all activities, a site map, the cause of spill, an estimate of the amount of PCB spilled and the concentration of PCB in the spilled material, particulars on the spill and spill site, cleanup activities, records of shipment and disposal of PCB-contaminated materials, and a detailed report of any sampling and analysis activities and results.

C. Training of Responsible Staff

The staff who are potentially or actually involved in a PCB spill cleanup need to know about regulatory requirements, company policies, procedures, and safety. Training sessions, training aids such as video tapes, written policies on procedures and safety (i.e., detailed policy, procedures, and safety manuals), staffing plans, and other organizational or educational methods should be used whenever practical to assure that the spill coordinators and cleanup crews from each PCB owner or operation know their responsibilities and options for future PCB spill response.

D. PCB Spill Cleanup Kit

In preparation for future PCB spills, a PCB spill cleanup kit should include the following (Phillips 1983):

- •55-gal. drums
- •Oil-absorbent material
- •Disposable gloves (several pairs)
- •Disposable plastic boots, jacket, pants, and plastic eye shields
- •Heavy-duty plastic sheeting to cover the spill site, buffer zone, storage areas, and the "cleaned" site after preliminary cleanup
- •Heavy tape, large plastic bags
- •Absorbent rags (5 lb)
- •Hand cleaner
- •Shovels, brooms, and dust pan
- •Solvent or cleaners
- •PCB signs, vinyl labels, rope, or other markers or barriers for site security
- •Phone numbers of National Response Center and appropriate company officials

Keep reasonable supplies of spill cleanup equipment on hand in each district or branch office that may supply a cleanup crew. Sampling equipment and materials are listed in the "Field Manual for Grid Sampling of PCB Spill Sites to Verify Cleanup" (Kelso et al. 1986).

E. Public Relations

A well-organized, efficient, and rapid response to a PCB spill tends to minimize public fear or concern about the spill. Communication with the public or media should be open. Describe the precautions taken to protect public health and indicate that the spill cleanup recommendations and standards of the EPA and other agencies are being followed. Disclosure of the results of verification analysis to interested parties also will reduce future concerns over the spill.

VI. References

Boomer BA, Erickson MD, Swanson SA, Kelso GL, Cox DC, Shultz BD. 1985. Verification of PCB spill cleanup by sampling and analysis. Report No. EPA-560/5-85-026, NTIS No. PB86-107315. Washington, DC: Office of Toxic Substances, U.S. Environmental Protection Agency.

Eisele PJ. 1985. PCB spill response and cleanup results. In Proceedings: 1985 EPRI PCB Seminar. Palo Alto, CA: Electric Power Research Institute. EPRI CS/EAEL-4480.

Gauger GA, Smith GC, Sullivan JM. 1985, Evaluation of a portable test kit for testing PCB contaminated soils and oils in the field. In Proceedings: 1985 EPRI PCB Seminar. Palo Alto, CA: Electric Power Research Institute. EPRI CS/EA/EL-4480.

Kelso GL, Erickson MD, Cox DC. 1986. Field manual for grid sampling of PCB spill sites to verify cleanup. Washington, DC: Office of Toxic Substances, U.S. Environmental Protection Agency. (In press.)

Lincoln Electric System. 1982. PCB spill reporting and cleanup. Procedure No. 108. Lincoln, NE. 12-01-82 revision.

NIOSH. 1977. National Institute for Occupational Safety and Health. Criteria for a recommended standard...occupational exposure to polychlorinated biphenyls (PCBs). U.S. Department of Health, Education, and Welfare (Public Health Service, Center for Disease Control, and National Institute for Occupational Safety and Health), DHEW (NIOSH) Publication No. 7-225.

NIOSH. 1980. National Institute for Occupational Safety and Health, U.S. Department of Health and Human Services. Health Hazard Evaluation Report No. 80-85-745. Oakland, CA: Pacific Gas and Electric Company.

Phillips BR. 1983. PCB: Electric utility system guide to use, marking, storage, record keeping, and disposal of PCBs and PCB items. Topeka, KS: Kansas Electric Cooperatives, Inc. 1985 edition.

USEPA. 1985. Polychlorinated biphenyls in electric transformers. Final Rule. (50 FR 29170-29201).

USEPA. 1982. PCB spill manual: prevention and emergency response.
Draft. Washington, DC: U.S. Environmental Protection Agency.
USWAG. 1984. The Utility Solid Waste Activities Group. Proposed spill
cleanup policy and supporting studies. Submitted to the U.S. Environmental
Protection Agency.

Index